Boris Gomelsky

Fish Genetics

Boris Gomelsky

Fish Genetics

Theory and Practice

VDM Verlag Dr. Müller

Impressum/Imprint (nur für Deutschland/ only for Germany)
Bibliografische Information der Deutschen Nationalbibliothek: Die Deutsche Nationalbibliothek
verzeichnet diese Publikation in der Deutschen Nationalbibliografie; detaillierte bibliografische
Daten sind im Internet über http://dnb.d-nb.de abrufbar.
Alle in diesem Buch genannten Marken und Produktnamen unterliegen warenzeichen-, marken-
oder patentrechtlichem Schutz bzw. sind Warenzeichen oder eingetragene Warenzeichen der
jeweiligen Inhaber. Die Wiedergabe von Marken, Produktnamen, Gebrauchsnamen,
Handelsnamen, Warenbezeichnungen u.s.w. in diesem Werk berechtigt auch ohne besondere
Kennzeichnung nicht zu der Annahme, dass solche Namen im Sinne der Warenzeichen- und
Markenschutzgesetzgebung als frei zu betrachten wären und daher von jedermann benutzt
werden dürften.

Coverbild: www.ingimage.com

Verlag: VDM Verlag Dr. Müller GmbH & Co. KG
Dudweiler Landstr. 99, 66123 Saarbrücken, Deutschland
Telefon +49 681 9100-698, Telefax +49 681 9100-988
Email: info@vdm-verlag.de

Herstellung in Deutschland:
Schaltungsdienst Lange o.H.G., Berlin
Books on Demand GmbH, Norderstedt
Reha GmbH, Saarbrücken
Amazon Distribution GmbH, Leipzig
ISBN: 978-3-639-32805-9

Imprint (only for USA, GB)
Bibliographic information published by the Deutsche Nationalbibliothek: The Deutsche
Nationalbibliothek lists this publication in the Deutsche Nationalbibliografie; detailed
bibliographic data are available in the Internet at http://dnb.d-nb.de.
Any brand names and product names mentioned in this book are subject to trademark, brand
or patent protection and are trademarks or registered trademarks of their respective holders. The
use of brand names, product names, common names, trade names, product descriptions etc.
even without a particular marking in this works is in no way to be construed to mean that such
names may be regarded as unrestricted in respect of trademark and brand protection legislation
and could thus be used by anyone.

Cover image: www.ingimage.com

Publisher: VDM Verlag Dr. Müller GmbH & Co. KG
Dudweiler Landstr. 99, 66123 Saarbrücken, Germany
Phone +49 681 9100-698, Fax +49 681 9100-988
Email: info@vdm-publishing.com

Printed in the U.S.A.
Printed in the U.K. by (see last page)
ISBN: 978-3-639-32805-9

To my family, friends and colleagues

Acknowledgements

I would like to acknowledge my colleagues and collaborators in the United States, Israel and Russia with whom I obtained data on fish genetics used in this book. My special thanks to Nina Cherfas, my teacher and long-term collaborator, and Robert Durborow for the help in preparation of this book.

I would like to acknowledge all authors whose data I mentioned and considered in this book and whose figures and tables have been reproduced. In order to follow a textbook format, I was obligated to omit formal citations for some presented materials.

Table of Content

Chapter 1. Fish Chromosomes and Cytogenetics of Fish Reproduction

1.1. Variability and Evolution of Fish Karyotypes

1.1.1. Introduction: Chromosomes and Karyotypes

In fish, as in the other eukaryotic organisms, genetic material is organized into structures called **chromosomes**. Chromosomes consist of DNA, macromolecule coding genetic information, and specific proteins called histones. Chromosomes change their structure depending on the phase of mitosis. Chromosomes reach maximum compactness at the stage of mitotic metaphase, when they are located at equatorial plane of division forming the metaphase plate. The counting of chromosomes and investigation of their structure are performed by analysis of metaphase plates.

Metaphase chromosomes consist of two chromatids joined by a **centromere**, a special structure, to which the mitotic spindle fibers attach. Parts of chromosomes separated by the centromere are named **chromosome arms**. Depending on the location of centromere the following types of chromosomes may be distinguished (Figure 1.1.1):

A - metacentric - the chromosome arms are equal;

B - submetacentric - the chromosome arms are different but this difference is not profound;

C - subtelocentric - the chromosome arms are very different;

D - acro(telo)centric - the centromere is located very close to one of the ends of the chromosome.

Somatic cells of organisms reproducing by a normal sexual mode have double or diploid (2n) set of chromosomes consisting of two haploid (n) sets. One haploid set has a maternal origin and the other, a paternal origin. Diploid set consists of pairs of **homologous chromosomes** having the same gene composition, sizes, and structure (besides the pair of sex chromosomes).

The set of chromosomes from somatic cell arranged according to their size and form is called **karyotype**. Two main indices are determined during karyological analysis: diploid number of chromosomes (designated as 2n), and the number of chromosome arms (designated as N.F., from Latin *Nombre Fondamental* - basic number). Metaphase chromosomes are systematized, i.e. they are placed according to their form and size.

Figure 1.1.1. Types on metaphase chromosomes based on location of centromere.

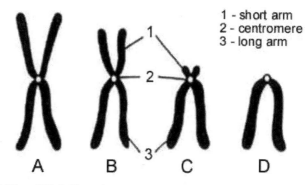

1 - short arm
2 - centromere
3 - long arm

A   B   C   D

1.1.2. Variability of Fish Karyotypes

Up to now karyotypes of about two thousand fish species have been described. Fish karyotypes are characterized by two specific features:

• High variability - diploid number of chromosomes varies in different species from 12 to 270;

• Sex chromosomes are usually not identified in karyotypes.

The observed high variability in karyotypes is explained by the fact that fishes are a very ancient and highly heterogeneous group of animals, evolution of which has taken place during several hundred millions years.

The distribution of fish according to diploid number of chromosomes (2n) is shown in Table 1.1.1. You can see in the table that more than 75% of fish have 42-60 chromosomes; from this class the most common diploid numbers are 48 and 50. It is supposed that these diploid numbers were typical for the ancestors of existing fish. During subsequent long-term evolution essential transformations of karyotypes occurred, which resulted in a decrease or increase in chromosome numbers.

Table 1.1.2 presents data on chromosome sets of aquaculture species and some fish inhabiting natural waters. Examples of karyotypes of several fish are presented in Figures 1.1.2, 1.1.3 and 1.1.4.

Table 1.1.1. Distribution of fish according to diploid number of chromosomes.

Diploid number of chromosomes, 2n	20 and less	22-40	42-60	62-80	82-100	102 and more
Frequency, %	0.7	11.8	76.3	4.0	4.8	2.4

Karyotype of grass carp (Figure 1.1.2) has 48 chromosomes; as was mentioned above, this number is very typical for fish karyotypes.

Karyotype of rainbow trout (Figure 1.1.3), consists of 44 metacentric and submetacentric chromosomes, 14 acrocentric chromosomes, and 2 subtelocentric chromosomes. Total number of chromosomes (2n) is 60. Another index of karyological analysis, N.F., is determined based on composition of chromosomes with different structure in karyotype. N.F. is calculated by assuming that each metacentric or submetacentric chromosome has 2 arms, while subtelocentic or acrocentric chromosome has 1 arm. Therefore, for this karyotype (Figure 1.1.3) N.F. is determined as 44 (number of meta- and submetacetric chromosomes) x 2 + 16 (number of acrocentric chromosomes) = 104.

In karyotypes of sturgeons some chromosomes have a very small size (Figure 1.1.4); they are called microchromosomes. The number of microchromosomes may vary; therefore for sturgeons the value of 2n is given with a range of variation (see Table 1.1.2).

Figure 1.1.2. Karyotype of grass carp (2n=48); chromosome pairs are numbered (from Beck et al. 1980; reproduced with permission from American Fisheries Society).

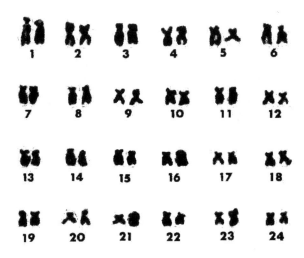

1.1.3. Evolution of Fish Karyotypes

During evolution the number of chromosomes may be changed by two main ways: 1) by **polyploidy**, i.e. by adding whole chromosome sets, and 2) by **chromosome rearrangements**.

The evidence that polyploidization played a significant role in fish evolution is occurrence of multiple differences in the chromosome numbers between fishes of close systematic groups. For example, most of fish in family *Cyprinidae* have in their karyotypes 48-50 chromosomes, whereas karyotypes of common carp and goldfish consist of about 100 chromosomes (see Table 1.1.2). This shows that during the evolution genomes of these two species passed through polyploidization, and common carp and goldfish may be regarded as tetraploids (4n) relative to most other cyprinid fish. It is supposed that this polyploidization occurred about 50 millions years ago.

Table 1.1.2. Data on chromosome sets in some fish species.

Fish family and species	2n	N.F.
Polyodontidae		
Paddlefish (*Polyodon spathula*)	120	160
Acipenseridae		
Shovelnose sturgeon (*Scaphirhynchus platorynchus*)	112	152
Beluda (Giant sturgeon) (*Huso huso*)	118±2	186
Sterlet (*Acipenser ruthenus*)	118±2	188-200
Sevruga (*Acipenser stellatus*)	118±2	188
Russian sturgeon (*Acipenser gueldenstaedtii*)	250±8	339-342
Siberian sturgeon (*Acipenser baeri*)	248±5	308
Lepisosteidae		
Longnose gar (*Lepisosteus osseus*)	56	90
Amiidae		
Bowfin (*Amia calva*)	46	66
Salmonidae		
Atlantic salmon (*Salmo salar*)	54-60	72-74
Brown trout (*Salmo trutta*)	78-82	98-100
Rainbow trout (*Oncorhynchus mykisss*)	58-62	104
Pink salmon (*Oncorhynchus gorbuscha*)	52-54	100-104
Sockeye salmon (*Oncorhynchus nerka*)	58	102-104
Chinook salmon (*Oncorhynchus tshawytscha*)	68	100-104
Coho salmon (*Oncorhynchus kisutch*)	60	100-104
Chum salmon (*Oncorhynchus keta*)	74	100-102
Esocidae		
Pike (*Esox lucius*)	50	50
Muskellunge (*Esox masquinongy*)	50	50
Cyprinidae		
Golden shiner (*Notemigonus crysoleucas*)	50	90
Fathead minnow (*Pimephales promelas*)	50	98
Grass carp (*Ctenopharyngodon idella*)	48	84-88
Silver carp (*Hypophthalmichthys molitrix*)	48	68-86
Bighead carp (*Hypophthalmichthys nobilis*)	48	74-86
Common carp (*Cyprinus carpio*)	100-104	148-160
Goldfish (*Carassius auratus auratus*)	100	148-154
Catostomidae		
Blue sucker (*Cycleptus elongatus*)	96-100	-
Bigmouth buffalo (*Ictiobus cyprinellus*)	100	-

Table 1.1.2. Continued

Ictaluridae		
Blue catfish (*Ictalurus furcatus*)	58	84
Channel catfish (*Ictalurus punctatus*)	58	92
Yellow bullhead (*Ameiurus natalis*)	60-62	82-84
Moronidae		
Striped bass (*Morone saxatilis*)	48	50
White bass (*Morone chrysops*)	48	-
Centrarchidae		
Bluegill (*Lepomis macrochirus*)	48	48
Green sunfish (*Lepomis cyanellus*)	46-48	48
Largemouth bass (*Micropterus salmoides*)	46	48
Smallmouth bass (*Micropterus dolomieu*)	46	48
Black crappie (*Pomoxis nigromaculatus*)	48	48
Percidae		
Yellow perch (*Perca flavescens*)	48	48
Walleye (*Sander vitreus*)	48	48
Cichlidae		
Nile tilapia (*Oreochromis niloticus*)	44	50
Mozambique tilapia (*Oreochromis mossambicus*)	44	46-50

A similar situation is observed among sturgeons (family *Acipenseridae*). Most fish from this family have about 118-120 chromosomes, while karyotypes of several species (such as Russian, Siberian and American green sturgeons) consist of 240-260 chromosomes (see Table 1.1.2 and Figure 1.1.4). It is supposed that there were two successive doublings of chromosome sets during evolution of sturgeons, and ancestors, which have not maintained up to now, had about 60 chromosomes.

Polyploidization is confirmed also by the analysis of DNA content in nuclei. The 2n and 4n cyprinid fish have about 2.0 and 4.0 picograms of DNA per nucleus, respectively. Among sturgeons the content of DNA in different species reflects multiplication of genomes also.

Fish from several other families also passed through doubling of their chromosome sets during evolution. For example, all fish from families *Catostomidae* (suckers) and *Salmonidae* (salmonids) are polyploids by their origin. This shows that polyploidy occurred repeatedly and independently in different systematic groups of fishes.

Figure 1.1.3. Karyotype of rainbow trout from strain "PdD 66" (2n=60; NF=104); M – metacentric, SM – submetacentric, ST – subtelocentric, A – acrocentric; (from Flajšans and Ráb 1990, Aquaculture 89:1-8; Copyright 1990, reproduced with permission from Elsevier).

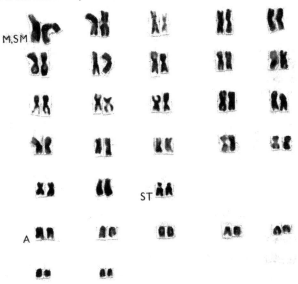

It needs to take into account that during long-term evolution the genome of ancient tetraploids passed through the process of secondary diploidization. Therefore all fish, which are tetraploid by their origin, are functional diploids. Nevertheless some genes (loci) remained in their duplicate state, and this duplication can be revealed by investigating of biochemical (protein polymorphism) or DNA genetic markers (Chapter 4).

During evolution karyotypes may be changed also by means of chromosome rearrangements. There are several types of chromosome rearrangements: **translocation** - exchange of regions within one chromosome or between different chromosomes; **inversion** - turning around by 180 degrees of chromosomal region; **duplication** - doubling of chromosomal region, and **deletions** - the loss of some chromosomal region. In some cases chromosome rearrangements can result in change in the number of chromosomes or chromosome arms.

Figure 1.1.4. Karyotype (2n=256) of American green sturgeon (*Acipenser medirostris*) (from Van Eenennaam et al. 1999; reproduced with permission from American Fisheries Society).

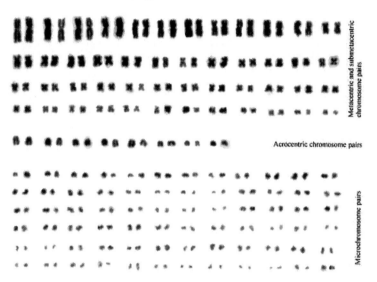

A special type of translocation, centric fusions, played a significant role in transformation of fish karyotypes. Centric fusion is joining of two acrocentric chromosomes in the centromere region with formation of one metacentric or submetacentric chromosome (Figure 1.1.5). As a result, the number of chromosomes decreased, but the number of chromosome arms did not change. This type of chromosome rearrangement is called **Robertsonian translocation** after the cytogeneticist who described it for the first time. Centric fusions had great significance in the evolution of karyotypes of some groups of fish. For example the numbers of chromosomes in salmons from the genera *Oncorhynchus* vary from 74 to 52, however the numbers of arms are practically equal (see Table 1.1.2). Theoretically the opposite process, centric fission, is also possible, but its probability is much lower, since in this case the appearance of an additional centromere is needed.

Figure 1.1.5. Scheme of formation one metacentric chromosome from two acrocentric chromosomes.

Chromosome rearrangements may also result to intraspecies variability of fish according to their karyotypes. For example, such variability is observed in salmonid fish: the diploid number of chromosomes in rainbow trout (2n) may vary from 58 to 62, and in brown trout from 78 to 82 (see Table 1.1.2).

Data of karyological analysis are used in evolutionary and taxonomic investigations of fish. The study of chromosome sets is needed also for the investigation of distant hybrids (Chapter 5) as well as for development of chromosome set manipulation methods such as induced polyploidy and induced gynogenesis (Chapter 6).

1.2. Sex Chromosomes and Genetics of Sex Determination

Sex of each organism is controlled by action of sex-determining genes, which switches the development program in either the "male" or "female" direction. In most animals the balance of sex-determining genes is regulated by the system of sex chromosomes.

Besides sex determination of individuals, the system of sex chromosomes also plays another role - it regulates sex ratio in progeny. One sex, **homogametic**, has two similar sex chromosomes, while another sex, **heterogametic**, has two different sex chromosomes. Heterogametic sex may be either male or female; in first case (male heterogamety) the sex chromosomes are usually designated as X and Y, in second case (female heterogamety) - as Z and W. These systems result in equal sex ratios in progenies (Figure 1.2.1).

In many organisms sex chromosomes are heteromorphic, i.e. they distinguish one from another according to their form and size. On this basis it is possible to identify them by means of karyological analysis. As mentioned in previous part (1.1), the sex chromosomes are not usually identified in fish karyotypes (isomorphic). Distinct sex chromosomes have been identified in approximately 10% of all karyologically investigated fish species (Devlin and Nagahama 2002). More frequently male heterogamety (XX - females, XY - males) and its modifications have been observed. Among aquaculture species male heterogamety was described in rainbow trout (Thorgaard 1977). In sockeye salmon the system of sex chromosomes XX-X0 has been revealed (Thorgaard 1978). Female karyotype has 58 chromosomes including two X-chromosomes, while male karyotype has 57 chromosomes with only one X-chromosome (and no Y-chromosome). Such a system is a modification of the classical XX-XY system; it is observed when the Y-chromosome is attached to one of the autosomes.

Figure 1.2.1. Scheme of sex determination in the case of male (A) and female (B) heterogamety.

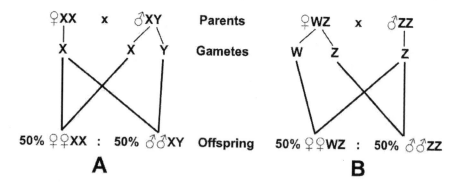

In other fish species the karyological investigations revealed female heterogamety (WZ- females, ZZ - males) and its modifications.

Even if sex chromosomes are not identified in karyotypes, their existence may be revealed genetically. The following two methods are usually used for the determination of the type of heterogamety in fish:

1. Induced gynogenesis. (Detailed description of this method is given in Chapter 6). In the case of female homogamety gynogenetic progenies are all-female, while in the case of female heterogamety they consist of both females and males. According to sex composition in gynogenetic progenies the female homogamety (and correspondingly male heterogamety; females - XX, males - XY) was revealed in common carp, grass carp, silver carp, rainbow trout, northern pike. The female heterogamety (females - WZ, males - ZZ) was shown for white sturgeon, plaice, muskellunge and some other species.

2. Crossing of sex-reversed fish. By means of hormonal treatment applied at early stages of ontogeny it is possible to induce hormonal phenotypic sex reversal (or inversion), i.e. the development of testes in genotypic females, or development of ovaries in genotypic males (see 1.3 and Chapter 6). However, the formula of sex chromosomes in reversed fish remains unchanged, and according to sex ratio in progenies obtained by crossing of such breeders it is possible to identify the type of heterogamety. In the case of male heterogamety the reversed males XX, when being crossed with normal females XX, produce all-female progeny:

$$♂_{rev.}\textbf{XX} \times ♀\textbf{XX} → 100\% ♀♀\textbf{XX}$$

In this way, female homogamety was revealed or confirmed in common carp, grass carp, rainbow trout, Atlantic salmon, crappie, yellow perch, walleye and some other fish species. In the case of female heterogamety crossing of sex reversed males WZ with normal females WZ gives the progeny with female-to-male ratio 3:1:

$$♂_{rev.}\textbf{WZ} \times ♀\textbf{WZ} → 25\% ♀♀\textbf{WW} : 50\% ♀♀\textbf{WZ} : 25\% ♂♂\textbf{ZZ}$$

In this way, female heterogamety was revealed in blue tilapia (*Oreochromis aureus*).

In summary, in many fish species, in which the sex chromosomes have not been revealed by karyological analysis, their existence has been proved genetically. Most frequently male heterogamety (males - XY; females - XX) was observed, but species with female heterogamety (females - WZ, males - ZZ) were revealed also. Apparently, most fish have cytologically identical (isomorphic) sex chromosomes. The isomorphism of sex chromosomes and the presence of both male and female heterogamety indicate to certain plasticity in sex-determining mechanisms in fishes. Sometimes different types of heterogamety were observed in very close species. For example, in genus *Esox* male heterogamety (XY/XX) was revealed in northern pike (*Esox lucius*) but female heterogamety (WZ/ZZ) was found in muskellunge (*Esox masquinongy*) (Dabrowski et al. 2000). Similarly, in tilapias (genus *Oreochromis*) Nile tilapia (*Oreochromis niloticus*) and Mozambique tilapia (*Oreochromis mossambicus*) have male heterogamety while female heterogamety was revealed in blue tilapia (*Oreochromis aureus*).

1.3. Formation of Reproductive System and Sex Differentiation

Development of reproductive system includes processes of gametogenesis and gonadogenesis. **Gametogenesis** is the process of transformation of primary sex cells to gametes – eggs and spermatozoa. **Gonadogenesis** is the process of formation and development of sex glands (gonads).

In fish, as in other vertebrate animals, the **primordial** (embryonic) **germ** (reproductive) **cells** (designated as **PGC**), the source of all sex cells of the adult organism, are separated at the very early stages of embryonic development. The PGC become morphologically distinguished at the stage of the late gastrula. Usually they have a much larger size than the cells surrounding them. Initially PGC appear in the marginal part of the embryo (ectomesoderm). They then migrate to a place of formation of the gonads, between the nephric tubules and the gut. In this location, at the dorsal side of body cavity, PGC line up in two longitudal strings. Later peritoneal epithelium (tissue lining a body cavity) protrudes into the body cavity and covers the primordial germ cells. As a result of this process gonads are formed.

The total number of PGC in fish embryos is not very large; usually it varies from 30 to 50. During migration PGC do not proliferate. Their proliferation begins after gonad formation. The sex cells, which appeared after mitotic divisions of PGC, are called gonia (singular - gonium). With numerous successive divisions the number of gonia increases rapidly. Simultaneously, somatic tissue of the gonads also develops. As a result of these processes, the gonads significantly increase their sizes.

The initial process of gonad development is the same in females and in males; therefore this period is called indifferent. The next stage of development is sex differentiation, i.e. appearance of traits typical of either males or females. Anatomical and cytological sex differentiations are separate processes. Anatomical differentiation includes the appearance of differences between males and females according to the gonad structure. Cytological differentiation results in the appearance of differences in development of sex cells. In genotypic males indifferent gonia transform to spermatogonia and the process of spermatogenesis begins. In genotypic females oogonia appear and the process of oogenesis begins.

As a rule, anatomical sex differentiation precedes cytological differentiation in fish. The ways of anatomical sex differentiation may vary in fish of different systematic groups. For example, in common carp (and other cyprinid fish) the first difference between testes and ovaries is the type of gonad attachment to the wall of the body cavity. The anatomical structures of ovary and testis of 3-month-old common carp are presented in Figure 1.3.1. The primordial ovary (Figure 1.3.1, A) is connected to the cavity wall in two points, by two mesovariums. The ovarian cavity is formed between the ovary and the wall of body cavity. The young testis (Figure 1.3.1, B) maintains the form of an elongated ridge attached to the body cavity wall in one point by an unpaired mesorchium.

The direction of sex differentiation is determined by the action of sex-determining genes, which start up the development program in the corresponding direction

Figure 1.3.1. Anatomic structure of gonads in 3-month-old common carp; A – ovary, B – testis; MO – mesovariums, OC – ovarian cavity, MCH – mesorchium.

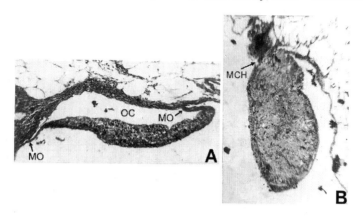

(female or male). This program is realized through a successive chain of hormonal regulations. The female steroid hormones (estrogens) in females and male steroid hormones (androgens) in males play a significant role at the initial stages of sex differentiation. On this basis, the artificial regulation of sex differentiation in fish by sex hormones is possible. The treatment of genotypic females with androgens may shift the normal process of sex differentiation and induce the development of testes. Similarly by estrogen treatment it is possible to induce the development of ovaries in genotypic males. Phenotypic sex transformation is called sex reversal (or sex inversion). The necessary condition for achievement of sex reversal is treatment during the period of sex differentiation. Currently hormonal sex reversal is widely used for artificial sex regulation in many aquaculture species (Chapter 6).

The possibility of hormonal sex reversal indicates to certain plasticity in the development and functioning of reproductive system in fish. Another obvious demonstration of this plasticity is the occurrence of hermaphroditism in fish. Hermaphroditism was described in approximately 20 fish families. Among invertebrates hermaphroditism is typical for primitive animals. Hermaphroditism in fish is secondary by its origin; it was evolved from gonohorism, i.e. from existence of two separate sexes. Hermaphroditism was found among evolutionary more advanced fish, such as sea basses (family *Serranidae*), sea breams (*Sparidae*), or wrasses (*Labridae*) (order *Perciformes*). Hermaphroditism in fish may be of two types - synchronous and consecutive (or successive). Synchronous hermaphrodites have male and female sexual cells simultaneously, and self-fertilization is possible. In the case of consecutive hermaphroditism the individual functions first as a female and afterwards as a male (protogyny), or in the opposite direction, from male to female (protandry). In case of hermaphroditism the action of genes determining the development of male and female traits is relatively equal. The program of development of these traits in ontogeny is determined by the interaction between genotype and environment. It is supposed that hermaphroditism in fish appeared as an adaptation for the optimal regulation of sex ratio in populations.

Usually some reptiles, e.g., crocodiles and turtles, are used as classical examples of so-called temperature dependent sex determination. Results of some experiments have revealed that extreme temperatures may affect the direction of sex differentiation in fish also. For example, according to the data by Baroiller et al. (1995) the rearing of Nile tilapia fry at temperatures of 34-36°C significantly increased the proportion of males (up to 70-90%). It is known, however, that this species has genetic sex determination with male heterogamety (males - XY, females - XX). The mechanism of temperature effect on sex differentiation in fish is not clear yet. Apparently, the extreme temperatures change the path in the metabolism of steroid hormones.

1.4. Gametogenesis and Fertilization

1.4.1. Introduction: Gametogenesis and Meiosis

Meiosis is a very important component of gametogenesis. Meiosis is a type of division, which occurs only in sex cells. Meiosis has two main functions: a) reduction of chromosome number from the diploid state (2n), typical for somatic cells, to the haploid state (n), typical for gametes, and b) recombination of genetic material. Reduction in chromosome number during meiosis results from two consecutive cell divisions with only one reduplication of chromosomes. The female and male gametes differ to a great extent both morphologically and functionally. Therefore the processes of **oogenesis** and **spermatogenesis** have essential differences, although the general scheme of meiosis is universal. Figure 1.4.1 presents the basic scheme of chromosome behavior during gametogenesis in animals.

1.4.2. Oogenesis

The development of female sex cells begins with mitotic divisions of oogonia; as a result their number significantly increases. In interphase before meiosis the replication of DNA occurs; each chromosome reduplicates with formation of two sister chromatids.

Female sex cells coming into meiosis are named **oocytes**. Until completion of the first meiotic division they are referred as **primary oocytes**. The process of oocyte development in fish is divided into four basic periods:

- **Chromatin-nucleolus stage**
- **Previtellogenic (small) growth**
- **Vitellogenic (big) growth**
- **Maturation**

The scheme of development of fish oocytes during chromatin-nucleolus and growth stages is given Figure 1.4.2. This scheme shows the correspondence between the stages of oogenesis and the phases of meiosis. During the chromatin-nucleolus stage of oogenesis complex transformations in the structure of chromosomes, caused by conjugation and crossing-over in prophase I, take place.

Prophase I is divided to several sub-phases. During the first sub-phase of prophase I, leptonema, thin chromosome threads are evenly distributed all over the nucleus (see Figure 1.4.2). During this period, preparation for further conjugation of homologous chromosomes occurs. Conjugation   of homologous chromosomes (or synapsis) takes place during the next stage – zygonema. For zygonema the specific arrangement of chromosomes is typical, on one side of the nucleus a dense

Figure 1.4.1. Scheme of chromosome behavior during gametogenesis in animals.

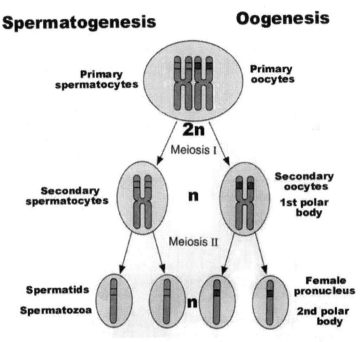

agglomeration of chromosomes is seen, while the other side of the nucleus is filled with single chromosome threads. This arrangement of chromatin distribution, sometimes called a "bouquet", is observed since conjugated chromosomes attach with their ends to a certain region of the nuclear membrane. At the end of zygonema chromosomes are condensed forming a one clew.

At the beginning of the pachynema, unwrapping of the chromosome clew begins. At the end of this stage thick chromosome threads are evenly distributed throughout the nucleus. At pachynema, homologous chromosomes tightly adjoin to each other; during this period the crossing-over occurs. At the beginning of diplonema homologous chromosomes begin to separate from each other and

Figure 1.4.2. Development of fish oocytes during chromatin-nucleolus and growth periods.

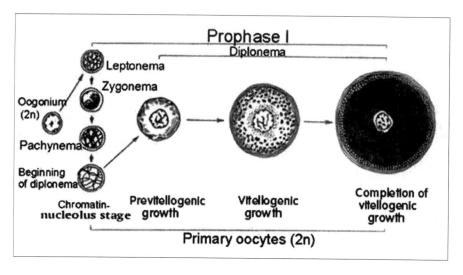

the space between them can be observed. By early diplonema the chromatin-nucleolus stage of development of the oocytes finishes, meiotic processes block and female sex cells begin to grow.

During the period of previtellogenic and vitellogenic growth, oocytes accumulate a great amount of nutritious substances needed for embryo development. During the growth period, high transcriptic activity of chromosomes, i.e. intensive synthesis and accumulation of ribonucleic acid (RNA), is observed. The storage of large amounts of RNA is necessary since the genes in fish embryos become transcriptionally active during gastrula stage. Until this period of development all proteins are synthesized on the basis of RNA accumulated previously in oocytes. For the synthesis of large quantities of RNA, numerous side loops, which become regions of active transcription, appear in chromosomes. Due to the appearance of such loops, chromosomes at this stage are named "lampbrush chromosomes". Numerous nucleoli are located at the edges of the oocyte nuclei during the period of growth; they also play a significant role in the RNA synthesis.

Since nucleoli are clearly visible during period of previtellogenic growth this stage is frequently referred as "perinucleolus stage".

During previtellogenic growth the cell size is increased by means of accumulation of homogeneous cytoplasm. Around the oocytes the follicle layer is formed, which is important for transfer of nutritional substances to oocytes. During vitellogenic growth oocytes accumulate yolk (or vitellus). Numerous vacuoles, oily drops and globules of yolk appear and gradually fill the whole cytoplasm. A multi-layer wall appears, surrounding the growing oocyte. The micropyle - a canal in the membranes through which a spermatozoon penetrates to the surface of the egg, is formed. During the period of vitellogenic growth oocytes drastically increase their volume and reach a final size. In the fully-grown oocytes the nucleus is located either in the middle of the cell or slightly shifted towards the animal pole, where the micropyle and a small region of cytoplasm free from yolk is located. At this stage the oocyte's large nucleus is frequently called the germinal vesicle (GV).

Oocyte growth may last many years depending on the age when females reach sexual maturity. During this period the oocytes are at the stage of diplonema of prophase I. Microphotographs of some stages in fish oocyte development are shown in Figure 1.4.3.

After completion of vitellogenic growth the oocytes are ready for transfer to the maturation stage. This transition is controlled by gonadotropic hormones, which are synthesized in pituitary gland (hypophysis). Under natural conditions hormone secretion is stimulated by spawning conditions. In aquaculture and fisheries artificial stimulation of maturation is frequently used. In this case females are injected with an extract of pituitary glands taken from other fish or with gonadotropic hormones of non-fish origin (or with synthetic analogs of gonadotropin-releasing hormones).

Oocytes that have completed the period of vitellogenic growth are at the stage of late diplonema of prophase I. At the transition to maturation the meiotic processes in female sex cells, which were blocked at the beginning of growth period, resume.

Figure 1.4.3. Microphotographs of common carp oocytes at different stages of development.

Oocytes at
chromatin-nucleolus stage

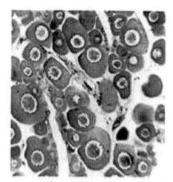

Oocytes of
previtellogenic growth

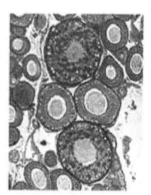

Oocytes of
vitellogenic growth

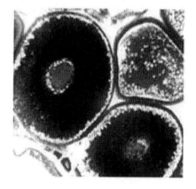

Oocytes completed
vitellogenic growth

The general scheme of nuclear transformations during the maturation period in fish oocytes is given in Figure 1.4.4.

The period of maturation begins with migration of the nucleus to the animal pole. Later, the nuclear membrane breaks down and chromosomes come out into the cytoplasm - at this period oocyte is at the final stage of prophase I – diakinesis (see Figure 1.4.4). Soon the spindle of the first meiotic division is formed, and chromosomes are placed in its equatorial plane forming the metaphase plate of metaphase I. After that the first meiotic division quickly completes. During the meiotic process in female sex cells only one egg results from one oogonium. The other products of meiotic divisions are called polar bodies, which later degenerate. In the first meiotic division the homologous chromosomes detach; therefore its products, the first polar body and secondary oocyte have haploid numbers of chromosomes (n), each of which consists of two chromatids (see Figure 1.4.1).

Figure 1.4.4. Nuclear transformations in oocytes during maturation period.

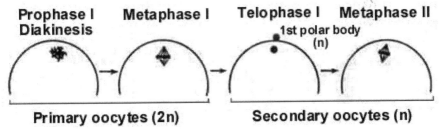

Almost at once after termination of the first meiotic division, the second division begins; it proceeds until metaphase II (See Figure 1.4.4). At this stage meiosis is blocked again, and oocytes ovulate: the follicle layer breaks down and oocytes come out to the ovarian cavity (in most fish) or in the body cavity (sturgeons and salmonids). Microphotograph of metaphase II in ovulated oocyte of silver crucian carp (*Carassius auratus gibelio, Cyprinidae*), is shown in Figure 1.4.5.

Ovulated oocytes, which are ready for fertilization, are called eggs; they can be spawned by females during natural spawning or stripped during artificial spawning in the hatchery. From a cytogenetic point of view the fish egg is a secondary oocyte at the metaphase of the second meiotic division. Second meiotic division terminates later during the fertilization process.

In contrast with a long growth period, the maturation of oocytes is a quick process. Depending on the temperature, the time period from the hormonal injection until the stripping of eggs lasts only about 12-24 hours.

Figure 1.4.5. Metaphase II in ovulated oocyte of silver crucian carp (courtesy of Nina Cherfas).

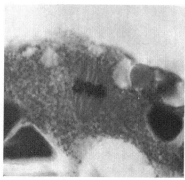

1.4.3. Spermatogenesis

The development of male sex cells begins with mitotic divisions of spermatogonia. As a result of their proliferation, **cysts** are formed – groups of cells originated from one spermatogonium and surrounded with a common capsule. Further development of male sex cells occurs in cysts located in the walls of seminiferous tubules. The cells in each cyst develop synchronously. The general scheme of spermatogenesis in fish is given in Figure 1.4.6.

The process of spermatogenesis may be divided to three periods:

- **Growth**
- **Maturation**
- **Formation**

The male sex cells coming into meiosis are called **primary spermatocytes**. During the growth period, transformations of chromosomes in primary spermatocytes are similar with those described above for oocytes at the chromatin-nucleolus stage. During this period some increase in size of sex cells is observed.

During the maturation period two consecutive meiotic divisions occur. In the first meiotic division the homologous chromosomes detach; therefore two daughter

Figure 1.4.6. Scheme of spermatogenesis in fish.

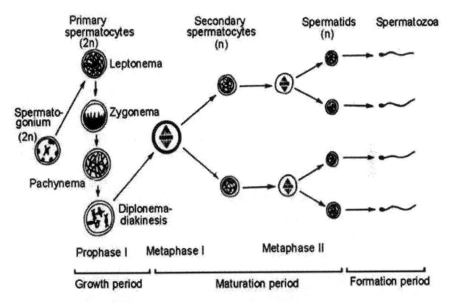

secondary spermatocytes are produced with haploid numbers of chromosomes, each of which consists of two chromatids (see Figures 1.4.1 and 1.4.6). The result of second meiotic division, in which sister chromatids detach, is that two spermatids are formed (see Figures 1.4.1 and 1.4.6).

During the period of formation the process of spermiogenesis occurs, i.e. transformation of spermatids to spermatozoa capable of active movement. After the completion of spermiogenesis the cyst capsules break out and spermatozoa are released to the lumens of seminiferous tubules.

During spermatogenesis from one primary spermatocyte four spermatozoa are formed, and a significant decrease in the size of male sex cells is observed. For example, in common carp (koi) the diameter of spermatogonia before the beginning of meiotic transformations is about 8 microns while the diameter of the spermatozoon head is only 1.5 microns.

Since neither nutritional substances nor products of transcription (RNA) are accumulated in male gametes, a delay of meiosis at the stage of prophase I in spermatogenesis does not occur. Besides this, in contrast with the eggs, spermatozoa are cells, which have fully completed meiosis. Therefore artificial stimulation of maturation in males by the injection of gonadotropic hormones is needed only to initiate secretion of spermatic fluid to dilute the agglomerations of spermatozoa.

Microphotograph of black crappie (*Pomoxis nigromaculatus*, family *Centrarchidae*) testis during the period of active spermatogenesis is given in Figure 1.4.7. You can see in the figure that cysts are located in walls of seminiferous tubules and cell size decreases during spermatogenesis.

Figure 1.4.7. Microphotograph of black crappie testis. PS - cyst with primary spermatocytes; SS - cyst with secondary spermotocytes; SD - cyst with spermatids; SZ - spermatozoa in the lumen of seminiferous tubule.

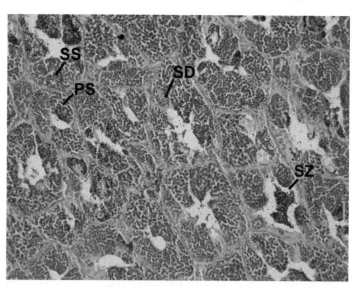

1.4.4. Fertilization

Fertilization is the process whereby male and female gametes (spermatozoon and egg, respectively) join to form a zygote, which is capable of developing into a new organism.

Fertilization is the first stage of embryonic development, which begins with insemination - the contact of a spematozoon (passed through micropylar canal) with the egg cytoplasm, and ending with karyogamy - the fusion of haploid nuclei of gametes resulting in the formation of a zygote with a diploid nucleus.

The scheme of nuclear transformations during the period of fertilization in fish is given in Figure 1.4.8.

Soon after penetration of the spermatozoon into the egg's cytoplasm, its head swells and transforms into the male pronucleus (see Figure 1.4.8). Simultaneously the second meiotic division in the egg is completed - the second polar body is extruded and the remaining haploid chromosome set transforms into the female

Figure 1.4.8. Nuclear transformations during fertilization period in fish.

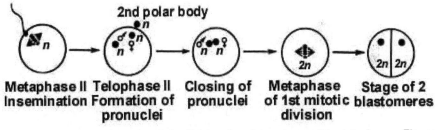

Metaphase II **Telophase II** **Closing of** **Metaphase** **Stage of 2**
Insemination **Formation of** **pronuclei** **of 1st mitotic** **blastomeres**
pronuclei **division**

pronucleus. (In the second meiotic division the sister chromatids detach, see Figure 1.4.1) The male and female pronuclei migrate into the depth of the cytoplasm and meet at the central part of the animal pole. During migration and closing of the pronuclei, the reduplication of chromosomes occurs with formation of two daughter chromatids. Later the membranes surrounding the pronuclei disappear and the male and female chromosomes fuse forming the metaphase plate of the first mitotic cleavage division. After its completion, two embryonic cells (blastomeres) are formed (see Figure 1.4.8). Microphotographs of some stages of nuclear transformations during period of fertilization are shown in Figure 1.4.9.

Figure 1.4.9. Microphotographs of some phases during fertilization period in silver crucian carp (*Carassius auratus gibelio*)*; A – extrusion of 2nd polar body; B – closing of pronuclei; C –metaphase of 1st cleavage division (courtesy of Nina Cherfas).

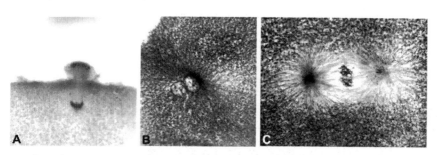

In order to observe nuclear transformations (forming and fusion of pronuclei, extrusion of 2nd polar body) the special histological treatment is needed. Under dissecting microscope it is possible to observe only formation of blastodisc (germinal disc), which appears as a result of concentration of cytoplasm at animal pole, and formation of two blastomeres in first cleavage mitotic division (Figure 1.4.10).

Figure 1.4.10. Early embryonic development in white bass x striped bass hybrids. A – formation of blactodisc at animal pole; B – stage of two blastomeres.

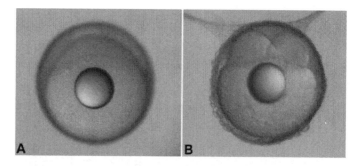

---

* This species has also all-female form which reproduces by natural gynogenesis (see 1.5). In this figure the nuclear transformations in form reproducing by means of normal sexual mode are shown.

The dynamics of nuclear transformations during fertilization in fish is determined by both characteristics of the species and temperature. In common carp at 20° C the extrusion of the second polar body is observed 11 min after insemination, and formation of two blastomeres – at 60 min. In rainbow trout at 10° C the same stages are observed at 70 and 450 min after insemination, respectively.

1.5. Natural Gynogenesis and Hybridogenesis

There are special types of sexual reproduction in animals, which are characterized by significant transformation of normal processes of meiosis and fertilization. The all-female forms reproducing by means of either gynogenesis or hybridogenesis are known among fish.

Gynogenesis (from Greek *gyno* - female, and *genesis* - origin) is a type of sexual reproduction where the male chromosomes are inactivated after insemination and embryonic development is controlled only by female chromosomes. In the case of gynogenesis the spermatozoon activates the egg to develop but the most important part of fertilization – karyogamy (fusion of nuclei) does not occur. Natural gynogenesis is a fairly rare event; it was revealed in about 10 fish species (forms) including the silver crucian carp (*Carassius auratus gibelio*), which is widely spread in Europe and Asia (Figure 1.5.1).

Figure 1.5.1. Silver crucian carp.

Gynogenesis in silver crucian carp was revealed in the middle 1940s by Russian geneticists K. Golovinskaya and D. Romashov. It was found that this species has two morphologically similar forms: (a) form consisting of females and males, which reproduces by the normal sexual mode, and (b) unisexual all-female form, which reproduces by means of gynogenesis. (Note that the other subspecies of crucian carp, *Carassius auratus auratus,* which also reproduces by the normal sexual mode, is the ancestor of goldfish.)

Differences in mode of reproduction were clearly revealed based on the results of crosses. The progeny from crossing females, reproducing by the normal sexual mode, with males of crucian carp consisted of females and males. When these crucian carp females were crossed with males of close species (e.g. common carp), hybrids were produced. Other results were observed after crossing of females from all-female form. In this case progenies consisted of females of crucian carp no matter what sperm (either from crucian carp males or males of other species) had been used for insemination of eggs. This proved that paternal chromosomes did not take part in embryo development, i.e. obtained progenies were of gynogenetic origin.

Karyological studies have shown that two forms of crucian carp differ according to the number of chromosomes – the fish reproducing in the normal sexual mode have 100 chromosomes, while fish reproducing by gynogenesis have about 150 chromosomes, i.e. gynogenetic form is triploid (3n):

Gynogenesis as a mode of reproduction has two main cytogenetic features:

- Inactivation of male chromosomes;
- Existence of special mechanisms preventing the reduction of chromosome number during meiosis in oocytes.

Non-reduction of chromosome number during oogenesis is an obligatory condition for the existence of gynogenetic forms, since it compensates for the absence of the male chromosome set.

In females of silver crucian carp reproducing by gynogenesis the constancy of chromosome number is reached by the radical transformation of meiosis in oocytes: the first meiotic division is absent, and the extrusion of the first polar body does not occur. Ovulated eggs are at metaphase of the only meiotic division and have an unreduced triploid (3n) chromosome number. After insemination, head of spermatozoon penetrated into the egg cytoplasm does not transform into the male pronucleus, maintains the appearance of a dense chromosomal clew, and is finally eliminated. At the same time, meiosis in eggs completes by extrusion of the only polar body. The group of chromosomes remaining in the cytoplasm transforms into a female pronucleus and later forms the metaphase plate of the first mitotic cleavage division.

Thus, whole meiotic process in gynogenetic females of silver crucian carp is reduced to one division, which is similar to usual mitosis. As a result, the recombination of genetic material, which is typical for normal meiosis, is completely eliminated. Therefore, gynogenetic offspring from each female have unchanged maternal genotype and are genetically identical. This was proven by experiments on tissue transplantation and later by DNA markers.

Natural gynogenesis in another fish – Amazon molly (*Poecilia formosa*, family *Poeciliidae*), which inhabits rivers flowing into the Gulf of Mexico, was described by American ichthyologists Carl Hubbs and Laura Hubbs in 1930s. This species is represented by only females. Amazon molly inhabit natural waters together with two related species, the sailfin molly (*Poecilia latipinna*) (Figure 1.5.2) and the Mexican molly (*Poecilia mexicana*), which reproduce by the normal sexual mode. (As you can see Amazon molly do not inhabit Amazon river. The name of fish is after Amazon warriors, a female-run society in Greek mythology.)

Females of the Amazon molly are inseminated by males of the two above mentioned species but the progenies consist of only females having the maternal genotype. It was found that in the natural populations, the Amazon molly is represented by two forms - diploid and triploid, and that each of them has a hybrid origin. The diploid form has 46 chromosomes; its karyotype consists of one haploid set of *Poecilia latipinna* (n=23) and one haploid set of *Poecilia mexicana* (n=23). The karyotype of the triploid form (3n=69) includes two haploid sets of *P. mexicana* and one haploid set of *P. latipinna*.

Figure 1.5.2. Female (left) and male (right) of sailfin molly (*Poecilia latipinna*). Males of this species inseminate females of Amazon molly.

Several other similar complexes consisting of several diploid species reproducing by means of the normal sexual mode, and all-female hybrid forms (frequently polyploid) reproducing by gynogenesis, have been described in fish. They have been found among minnows (genus *Phoxinus*), silversides (*Menidia*), topminnows (*Poeciliopsis*), spined loaches (*Cobitis*) and some other fish. The existence of such complexes shows the certain connection between gynogenesis, interspecies hybridization and polyploidy in fish. (Some considerations on origin of gynogenetic forms in fish are given in 5.4.2.)

The other type of sexual reproduction, **hybridogenesis**, has been found only in some topminnows (*Poeciliopsis*) and was described by R. Schultz in 1969. Hybridogenous forms are presented by only females that reproduce with the males of close species. Somatic cells of these females are diploid and consist of two haploid chromosome sets of different species. At early stages of oogenesis in females, all chromosomes of paternal origin are eliminated and only a haploid set of maternal chromosomes comes into the eggs. The normal process of fertilization occurs and, after fusion of pronuclei, the diploid hybrid individuals appear again. In the case of hybridogenesis the processes of elimination of male chromosomes and their contribution repeat in every new generation.

Chapter 2. Inheritance of Qualitative Morphological Traits in Fish

2.1. Introduction: Main Features of Qualitative Traits

All morphological traits may be divided to two large groups according to the character of their inheritance: **qualitative** and **quantitative**.

The qualitative traits have the following main features:

- Have alternative variability, which manifests through appearance of discrete phenotypic classes;
- Are determined by a relatively low number of genes;
- Their manifestation usually does not depend on environmental conditions;
- They are inherited according to Mendel's principles.

In fish, variability in qualitative traits results in the appearance of alternative types of color, scale cover, fin length etc.

2.2. Inheritance of Color Traits in Fish

2.2.1. Basic Principles of Color Development and Inheritance

The color of skin in fish is determined by the combination of color pigments. There are several types of specialized pigment-containing cells (chromatophores) in fish skin. Each type of chromatophores contains a certain kind of pigment. **Melanophores** contain the black pigment melanin. **Erythrophores** and **xanthophores** accumulate red and yellow pigments, respectively. **Iridophores** contain crystals of colorless pigment guanine, which refracts and reflects light giving fish their typical metallic appearance.

In fish, as in other animals, the hereditary variability in body color results from mutations of genes controlling the synthesis of pigments or the structure and distribution of pigment cells. The absence, decreased or increased amount of some pigment in the skin result in changes in body color, i.e. appearance of color morphs.

Numerous color variants have been described in ornamental fish cultivated in aquaria. For ornamental species such as guppy, goldfish, medaka, and angelfish, the genetic basis for color variability has been revealed and many color-modifying genes have been described. For aquaculture species and fish inhabiting natural waters the information on color variability is much more scarce. Color modifications have been revealed only in a few species. The appearance of albino individuals in many fish species may be regarded as an exception.

In the following parts of this chapter the data on color variability and inheritance in cultivated fish species (besides aquarium fish) are presented. Color modifications in fish are typical qualitative traits. They are inherited according to Mendel's principles and may be unraveled on the basic rules of monohybrid and dihybrid crosses. The information on different types of gene interaction such as complementary gene action, recessive and dominant epistasis, and action of duplicate genes is also essential for understanding of these data.

2.2.2. Albinism and Its Inheritance

The most frequent color modification observed in fish is **albinism**. Albinism is the absence of black pigment, melanin, both in the skin and in the eyes. The albino animals have a yellowish body and pink eyes. The eyes look pink or red since the blood vessels and capillaries in the iris and retina become visible due to the absence of melanin.

Albino mutants have been found among fish of very different systematic groups: lampreys, sharks, sturgeons and many teleost fish. Albino individuals have been described in many aquarium fish (Figure 2.2.1), several aquaculture species including rainbow trout, channel catfish (Figure 2.2.2), grass carp, and among fish inhabiting natural waters.

The inheritance of albinism has been analyzed in many aquarium fish and several aquaculture species - rainbow trout, channel catfish and grass carp. In all cases the albinism was controlled by autosomal recessive mutation. The crossings of parental wild-type color fish (*AA*) with albino fish (*aa*) results in appearance of wild-type color fish in F_1:

P: ♀ *AA* (wild-type) x ♂ *aa* (albino)

F_1:   *Aa* (all wild-type)

The crosses between F_1 fish give the classical Mendelian ratio phenotypic ratio 3:1 (wild type : albino):

F_1 x F_1:   ♀ *Aa* (wild-type) x ♂ *Aa* (wild-type)

F_2: 1 *AA* (wild-type) : 2 *Aa* (wild-type) : 1 *aa* (albino)

Figure 2.2.1. Albino corydoras catfish.

Bridges and von Limbah (1972), in a study on inheritance of albinism in rainbow trout, produced 22,535 fish in F_2. From them, 16,856 (74.8%) were of the wild type color and 5,679 (25.2%) had albino coloration.

In test-crosses of F_1 fish with albino fish the classical ratio of 1:1 was observed:

Aa (wild-type) x *aa* (albino) → 1 *Aa* (wild type) : 1 *aa* (albino)

For example, Rothbard and Wohlfarth (1993), in a study on inheritance of albinism in grass carp, have crossed an F_1 wild-type color male with an albino female. In the resulting progeny among 1020 fish, 530 (52.0%) fish were of wild-type color and 490 (48.0%) were albino.

As you will see in part 2.2.3, the albino mutant gene can interact with other color modifying genes. In these cases the color is determined according to the rules of dihybrid cross (modifications of 9:3:3:1 ratio in F_2).

Figure 2.2.2. Wild-type color (above) and albino (below) channel catfish.

Synthesis of melanin from the amino acid tyrosine is a complex metabolic process. It is a chain of biochemical reactions with formation of the next compound at each stage (A→B→C); each reaction needs the presence of specific enzyme. According to the admitted hypothesis "one gene – one enzyme", the synthesis of each enzyme is controlled by a separate gene. Therefore the whole process of melanin production is under control of several genes. Mutation of any of these genes can lead to termination of melanin synthesis and appearance of albino animals. So, theoretically albino individuals found in different populations of the same species may be products of mutation either of the same gene or two different genes. Thorgaard et al. (1995) have crossed albino rainbow trout originated from 6 different locations. When fish from 5 lines were crossed, the progenies consisted of albino fish only. But when fish of one strain (Chelan) were crossed with fish of other lines, pigmented wild-type color fish have been produced. Appearance of wild-type color fish indicated that the Chelan albinos resulted from mutation of another gene. This cross may be presented as:

$a^1a^1A^2A^2$ (albino/Chelan line) x $A^1A^1a^2a^2$ (albino/other line) → $A^1a^1A^2a^2$ (wild-type)

where A^1 and A^2, and a^1 and a^2 are dominant alleles of wild-type color and recessive mutant alleles of two genes, respectively. This type of cross of mutant animals, performed to determine whether one or two genes are involved, is called the complementation test. The appearance of wild-type animals in the resulting progeny indicates that these morphs are resulted from recessive mutations of two different genes.

The black pigment melanin has a photoprotective effect. It absorbs ultraviolet irradiation and, as a result, protects skin, eyes and whole body from damage. The lack of melanin in albino animals makes them more sensitive to visible light. In natural waters albino fish become more vulnerable for predators due to their light color. The growth rate and survival of albino fish are usually much less than those of normally pigmented fish. Therefore fish of this color modification are not commercially reared for food. Nevertheless albino mutants of some aquaculture species, such as grass carp, are cultivated as ornamental fish.

2.2.3. Color Modifications in Some Aquaculture and Sport Fish Species
2.2.3.1. Rainbow Trout

Several color modifications have been described in rainbow trout. Golden and palomino variants of rainbow trout have been identified in the United States. Body color of golden rainbow trout is similar to the albino morph except that eyes are pigmented. Body color of palomino trout is intermediate between golden and wild type fish. Golden and palomino color morphs result from absence or reduction of black pigment melanin in the skin. Golden and palomino trout are popular sport fish in West Virginia and Pennsylvania. According to the records (Clark 1971) these variants originated from one golden mutant rainbow trout female (found in a fish hatchery in West Virginia in 1956), which was crossed with a normal male. Later, the inheritance of these color modifications was revealed (Wright 1972). These color modifications in rainbow trout are caused by the action of a mutant allele G', which demonstrates incomplete dominance relative to the allele for wild color G. Each possible genotype results in the appearance of a specific phenotype:

<u>Genotype</u>	<u>Phenotype</u>
GG	**Wild type color**
G'G'	**Golden**
G'G	**Palomino**

The mutant allele G' induces a reduction in the amount of melanin in skin, the rate of this reduction depends on the number of mutant alleles in the genotype: heterozygotes (*G'G*) with palomino color have intermediate reduction of melanin while golden fish (*G'G'*) do not have black pigment in the skin at all.

Crossing of golden fish gives progeny consisting of only golden fish:

$G'G'$ (golden) x $G'G'$ (golden) → 100% $G'G'$ (golden)

Crossing of golden fish with wild type color fish resulted in F_1 progeny consisting of palomino fish (G'G). Crossing of F_1 fish resulted in a 1:2:1 ratio in F_2 progeny:

P: $G'G'$ (golden) x GG (wild type)

F_1: $G'G$ (palomino)

F_1 x F_1: $G'G$ x $G'G$

F_2: 1 $G'G'$ (golden) : 2 $G'G$ (palomino) : 1 GG (wild type)

In populations of rainbow trout cultivated in France another mutation of yellow (golden) color was revealed (Chourrout 1982). In contrast to the mutation G' described above with incomplete dominance, this mutant allele of yellow color is completely dominant relative to the allele of wild type color. This shows that similar color modifications may be caused by different mutations.

Dobosz et al. (1999) have shown that the appearance of albino and darker brownish-yellow color with pigmented eyes (also called by authors as "palomino") in rainbow trout cultivated in Poland is determined by interaction of two genes having two alleles each: A/a, and B/b. Recessive allele a is typical recessive mutation of albino color, but in this case its expression is modified by the other gene B/b. In fish having two recessive alleles a (genotype aa) in the presence of dominant allele B (genotypes $aaBB$ or $aaBb$) brown-yellow color is developed. The scheme and results of crosses are presented in Figure 2.2.3.

As you can see in Figure 2.2.3 only fish with genotype $aabb$ (1/16 of all fish) develop albino color in F_2. Resulting phenotype ratio 12:3:1 is a modification of the classical Mendelian ratio for a dihybrid cross 9:3:3:1. This ratio is typical for dominant epistasis when dominant allele of one gene (in this case dominant allele A) masks expression of another gene (B/b). At the early development stage it is impossible to distinguish between future brown-yellow and albino embryos. All embryos have light yellow body and pink eyes, and the original segregation 3:1 according to alleles A/a is recorded (shown in Punnett square in the figure with thicker lines). During further development and growth of fish, dominant allele B induces appearance of late pigmentation and phenotypic class of brown-yellow fish appears.

2.2.3.2. Tilapias

Apparently tilapias (genus Oreochromis) are the only group of fish cultivated for food, for which body color has an essential economic significance. This became obvious after the first appearance of Taiwanese red tilapia at the end of the 1970s. This strain had a hybrid origin and was developed in Taiwan by a hybridization of several (at least 3) species, among which Mozambique tilapia (Oreochromis mossambicus) and Nile tilapia (O. niloticus) predominated. This hybrid strain of red tilapia has been cultivated intensively, initially, in Taiwan and, later, in many regions of the word. Reddish color made these fish more attractive for customers in some countries of Asia and South America as compared to normally pigmented fish. This strain also had a good growth rate. Later strains of red tilapias were developed in different tilapia species. Other color modifications (besides red) in tilapias have also been described.

Figure 2.2.3. Inheritance of brown-yellow and albino colors in rainbow trout (dominant epistasis).

P: ♀ *AABB* (wild-type) x ♂ *aabb* (albino)

F$_1$: *AaBb* (wild-type)

F$_1$ x F$_1$: ♀ *AaBb* x ♂*AaBb*

G A M E T E S

♀ \ ♂	AB	Ab	aB	ab
AB	AABB wild-type	AABb wild-type	AaBB wild-type	AaBb wild-type
Ab	AABb wild-type	AAbb wild-type	AaBb wild-type	Aabb wild-type
aB	AaBB wild-type	AaBb wild-type	aaBB brown-yellow	aaBb brown-yellow
ab	AaBb wild-type	Aabb wild-type	aaBb brown-yellow	aabb albino

F$_2$: 12 wild-type : 3 brown-yellow : 1 albino

Color inheritance in Taiwanese red tilapia has been investigated for a long time, but the obtained data often have been contradictory and difficult to understand. Apparently it was connected with the hybrid nature of this strain. According to Wohlfarth et al. (1990) the red color in Taiwanese tilapia is determined by one gene P with two alleles P_1 and P_2 showing incomplete dominance. The homozygotes P_1P_1 have wild-type black color, homozygotes P_2P_2 are pink, while heterozygotes P_1P_2 are red. Sometimes homozygotes P_2P_2 had decreased survival and this made analysis of obtained ratios more complicated.

The data on inheritance of red color morphs of tilapia have been obtained by investigation of pure species also. McAndrew et al. (1988) have investigated the inheritance of several color modifications in Nile tilapia. It was revealed that red color in this species is caused by the dominant allele of one gene (R/r). The crossing of red fish RR with wild type color fish rr gave red F_1 heterozygotes Rr. The crossing of F_1 fish resulted in the classical 3:1 ratio (red : wild type color) in the F_2. Histological investigation of skin has shown that appearance of red color was caused by the absence of black pigment melanin in skin.

Wolhfarth et al. (1990) and Reich et al. (1990) revealed that in another species, Mozambique tilapia, in contrast to Nile tilapia, the red color is controlled by recessive allele. The red Mozambique tilapia had genotype bb, normal black color fish had genotypes BB or Bb. These data showed that the similar color morphs in tilapias can result from mutations of different genes, and mutant alleles may be either dominant or recessive relative to corresponding alleles of wild-type color.

Investigating the inheritance of red color in the Philippine hybrid strain of tilapia Reich et al. (1990) have described another color morph - bright color (or "albino with black eyes"). This color was controlled by a recessive mutation of gene I/i changing the structure of iridophores. The fish with genotype ii developed such color only in the presence in the genome of a dominant allele R of other gene, inducing by itself the development of red color. Fish with genotype rr have black wild-type color irrespective of I/i gene. Such type of gene interaction is called

recessive epistasis. The scheme and results of corresponding crosses are presented in Figure 2.2.4. The ratio 9:3:4 in F_2 is typical for recessive epistasis.

Figure 2.2.4. **Inheritance of red and bright colors in tilapia (recessive epistasis).**

P: ♀ *RRII* (red) x ♂ *rrii* (black)

F_1: *Rrli* (red)

F_1 x F_1: ♀ *Rrli* x ♂ *Rrli*

GAMETES

♂ / ♀	RI	Ri	rl	ri
RI	*RRII* red	*RRIi* red	*RrII* red	*Rrli* red
Ri	*RRIi* red	*RRii* bright color	*Rrli* red	*Rrii* bright color
rl	*RrII* red	*Rrli* red	*rrII* black	*rrli* black
ri	*Rrli* red	*Rrii* bright color	*rrli* black	*rrii* black

F_2: 9 red : 3 bright color : 4 black

Lutz (1999) described another example of color determination in tilapia by interaction of two different genes. It appeared that the light color in Nile tilapia (morph called "Nile Pearl", Figure 2.2.5) is developed when fish genotype contains at least one dominant allele of each of two genes (A and B). Such type of gene interaction is called **complimentary gene action**. The scheme and results of corresponding crosses are presented in Figure 2.2.6. The phenotypic ratio 9:7 in F_2 is typical for complementary gene action.

Figure 2.2.5. Nile Tilapia of light color ("Nile pearl").

2.2.3.3. Koi

The Japanese color carps or koi are popular objects of decorative aquaculture in many countries. Koi belong to the same species (*Cyprinus carpio*) as edible carp. Koi have been developed in Japan. The first color morphs were found in the early 1800s among a normal stock of common carp reared for food in small terrace ponds in Niigata province (Figure 2.2.7).

Now about 15 main color types of koi are distinguished. Most of koi color types are not true breeds, i.e. if fish of the same type are crossed, the progeny will be variable and will contain fish of different color types.

Katasonov (1978) has shown that melanin formation in wild-type color carp is controlled by two dominant duplicate (i.e. having similar action) dominant genes B_1 and B_2. When wild-type color common carp (genotype $B_1B_1B_2B_2$) (Figure 2.2.8.A) is crossed with koi (genotype $b_1b_1b_2b_2$), for example with white-red koi (Kohaku according to Japanese terminology (Figure 2.2.8.B), F_1 progeny consist of wild-type color fish only (genotype $B_1b_1B_2b_2$). When fish from F_1 are crossed, in F_2 progeny the phenotypic ratio 15:1 is observed. The scheme and results of these crosses are presented in Figure 2.2.9. The ratio 15:1 in F_2 is typical for interaction of duplicate genes.

Figure 2.2.6. Inheritance of light color in Nile tilapia (complimentary gene action).

P: ♀ *AABB* (light color) x ♂ aabb (wild-type)

F₁: AaBb (light color)

F₁ x F₁: ♀ AaBb x ♂AaBb

GAMETES

♀ \ ♂	AB	Ab	aB	ab
AB	AABB light color	AABb light color	AaBB light color	AaBb light color
Ab	AABb light color	AAbb wild-type	AaBb light color	Aabb wild-type
aB	AaBB light color	AaBb light color	aaBB wild-type	aaBb wild type
ab	AaBb light color	Aabb wild-type	aaBb wild type	aabb wild type

F₂: 9 (light color) : 7 (wild-type)

Presence of melanin in wild-type color fish is visible already at late embryonic and larval stages while koi embryos and larvae are transparent (Figure 2.2.10).

Figure 2.2.7. Scene of Niigata province - homeland of koi (photo by the author).

The information concerning inheritance of multicolor traits in koi is scarce. Gomelsky et al. (1996, 1998) investigated the color ratios in progenies obtained after crossing of two-color (white-red or white-black) and tri-color (white-red-black) fish. Photographs of some koi breeders, which were used in these studies, are presented in the Figure 2.2.11. Results of these studies have shown that the white-red color complex and the presence of black patches in koi are inherited independently. The

Figure 2.2.8. Common carp (wild-type) (A) and white-red (Kohaku) koi (B).

Figure 2.2.9. Inheritance of melanin formation in common carp (duplicate genes action).

P: ♀ $B_1B_1B_2B_2$ (wild type) x ♂ $b_1b_1b_2b_2$ (koi)

F$_1$: $B_1b_1B_2b_2$ (wild type)

F$_1$ x F$_1$: ♀ : $B_1b_1B_2b_2$ x ♂: $B_1b_1B_2b_2$

GAMETES

♀ \ ♂	B_1B_2	B_1b_2	b_1B_2	b_1b_2
B_1B_2	$B_1B_1B_2B_2$ wild type	$B_1B_1B_2b_2$ wild type	$B_1b_1B_2B_2$ wild type	$B_1b_1B_2b_2$ wild type
B_1b_2	$B_1B_1B_2b_2$ wild type	$B_1B_1b_2b_2$ wild type	$B_1b_1B_2b_2$ wild type	$B_1b_1b_2b_2$ wild-type
b_1B_2	$B_1b_1B_2B_2$ wild type	$B_1b_1B_2b_2$ wild type	$b_1b_1B_2B_2$ wild-type	$b_1b_1B_2b_2$ wild type
b_1b_2	$B_1b_1B_2b_2$ wild type	$B_1b_1b_2b_2$ wild-type	$b_1b_1B_2b_2$ wild type	$b_1b_1b_2b_2$ koi

F$_2$: 15 wild-type : 1 koi

presence of black patches is controlled by the dominant mutation of one gene (*Bl/bl*). Fish with genotypes *BlBl* or *Blbl* are white-black, red-black or white-red-black. Homozygotes *blbl* do not have black patches and may be white, red or white-red. When heterozygos (*Blbl*) white-red-black fish are crossed, all six possible color combinations are present in progeny (solid white, white-red, solid red, white-black, white-red-black and red-black) and the ratio of fish with black patches to fish without black patches is 3:1. The major gene *Bl/bl* determines only the presence or absence of black patches. The rate of development of black patches in fish having the *Bl* allele in their genotypes is highly variable - from a few dispersed black spots to large black patches covering large areas of the body. Apparently the rate of development of black patches is under the control of many genes. Obviously the same is true for the rate of fish coverage with red patches, which is also very variable.

Figure 2.2.10. Transparent (unpigmented) koi larvae (A) and pigmented (with melanin) common carp larvae (B).

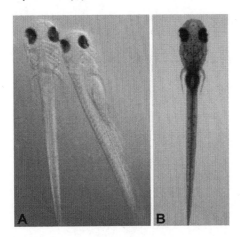

2.2.3.4. Crappie

Black crappie *Pomoxis nigromaculatus* is a popular sport fish in the United States. Occurrence of a specific color morph, fish having a distinctive black predorsal stripe, has been found in this species (Figure 2.2.12).

Gomelsky et al. (2005) have studied the inheritance of this trait. In the first generation produced by crossing fish with and without stripe all fish had black stripe. In the second generation the classical Mendelian ratio 3:1 was observed. Based on these data it was concluded that the appearance of the predorsal black stripe in crappie is under control of a dominant mutation of one gene (*St/st*). Fish with genotypes *StSt* and *Stst* have the predorsal stripe, while fish with genotype *stst* do not develop this trait.

Figure 2.2.11. Multi-color koi parents used for investigation of color inheritance; A – white-red fish (Kohaku); B – white-black fish (Shirro-Bekko); C – white-red-black fish (Sanke); D – white-red-black with round red spot on the head (Tancho Sanke).

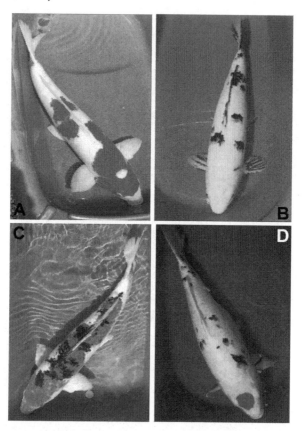

2.3. Inheritance of Scale Cover in Common Carp and Koi

The most well known example of heritable morphological variability among aquaculture species is the existence of several types of scale cover in common carp. Common carp (*Cyprinus carpio*) is the main aquaculture species in many countries of Europe and Asia. The same scale cover types are known in koi since koi is the ornamental form of the same species.

Figure 2.2.12. Black crappie with (right) and without (left) predorsal black stripe.

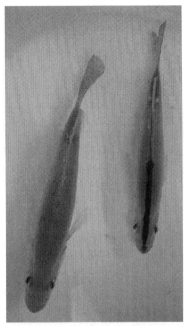

There are four main types of scale cover in common carp and koi. In **scaled** fish (wild type) (Figure 2.3.1.A) even rows of scales cover the whole body. **Mirror** (or scattered) fish have large ("mirror") scales which are scattered on the body (Figure 2.3.1.B). The mirror carp demonstrate high variability with regard to the reduction of scale cover. Some mirror fish have the body almost completely covered with big scales. In **linear** fish large scales form even row along the lateral line (Figure 2.3.1.C). In general, there are no scales on body of **leather** fish (Figure 2.3.1.D); several small scales may be found sometimes near the base of fins.

Data on the inheritance of scale cover in common carp were obtained by Russian scientists Kirpichnikov and Balkashina in the middle of the 1930s. The type of scale cover is determined by an interaction of two genes having two alleles each (*S/s* and *N/n*). Fish of different scale cover types have the following genotypes:

- Scaled - **SSnn** or **Ssnn**
- Mirror – **ssnn**
- Linear - **SSNn** or **SsNn**
- Leather – **ssNn**

Fish having the dominant allele N in the homozygote state (genotype NN) are unviable and perish at the time of hatching. Therefore 1/4 of progeny in crosses between linear and leather carps (heterozygotes Nn) are inviable offspring. The expected phenotypic ratios in all possible crosses are presented in Table 2.3.1. The crosses 1 and 2 (see Table 2.3.1) of scaled and mirror fish give ratios typical for monohybrid crosses since only one gene (S/s) is involved. In crosses 3, 4, 6 and 7 the resulting ratios are determined by allele combination of two genes, and all resulting offspring are viable. Crossing of mirror carps or koi (cross 5) gives only mirror fish. In crosses 8-10 between linear and leather fish (heterozygotes Nn) the inviable offspring with genotype NN appear (1/4 of progeny) (see Table 2.3.1).

Punnett square for the cross of two double heterozygotes (genotype $SsNn$, linear fish) is presented in Figure 2.3.2. Resulting phenotypic ratio among viable fish is 6:3:2:1 or 50% : 25% : 16.7% : 8.3%. This is a modification of the classical ratio 9:3:3:1, which is caused by inviability of fish with NN genotype.

Common carp with reduced scale cover have been known in Europe from the 17th century. It is supposed that, initially, mirror fish have been appeared as a result of a mutation $S \to s$. Later, another independent mutation $n \to N$ resulted in the appearance of linear and leather fish.

The genes for scale cover have wide pleiotropic effect, i.e. they influence many traits. The effect of allele N is especially strong. Linear and leather carps (genotype Nn) have retarded growth rate and decreased survival as compared with scaled and mirror carps (genotype nn). These differences become more pronounced, when fish are reared under unfavorable conditions. The pleiotropic effect of allele s is much weaker although some difference in growth rate between scaled and mirror fish is observed.

Figure 2.3.1. Different types of scale cover in koi; A – scaled fish, B – mirror fish, C – linear fish, D – leather fish.

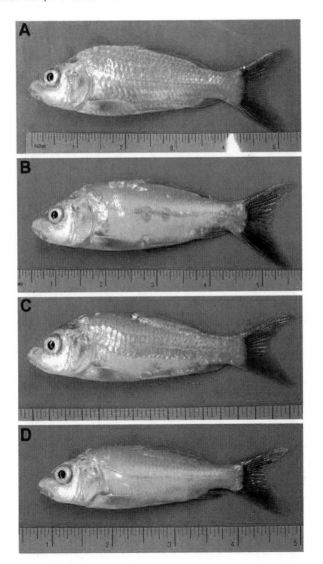

Table 2.3.1. Expected phenotypic ratios in crosses of common carp or koi with different scale cover types.

Parents and type of crossing	Phenotypes (%)			
	Scaled	Mirror	Linear	Leather
1. Scaled x Scaled:				
SSnn x *SSnn, SSnn* x *Ssnn*	100	0	0	0
Ssnn x *Ssnn*	75	25	0	0
2. Scaled x Mirror:				
SSnn x *ssnn*	100	0	0	0
Ssnn x *ssnn*	50	50	0	0
3. Scaled x Linear:				
SSnn x *SSNn, SSnnxSsNn* *Ssnn* x *SSNn*	50	0	50	0
Ssnn x *SsNn*	37.5	12.5	37.5	12.5
4. Scaled x Leather:				
SSnn x *ssNn*	50	0	50	0
Ssnn x *ssNn*	25	25	25	25
5. Mirror x Mirror:				
ssnn x *ssnn*	0	100	0	0
6. Mirror x Linear:				
ssnn x *SSNn*	50	0	50	0
ssnn x *SsNn*	25	25	25	25
7. Mirror x Leather:				
ssnn x *ssNn*	0	50	0	50
8. Linear x Linear:*				
SSNn x *SSNn, SSNn* x *SsNn*	33.3	0	33.3	0
SsNn x *SsNn*	25	8.3	50	16.7
9. Linear x leather:*				
SSNn x *ssNn*	33.3	0	66.7	0
SsNn x *ssNn*	16.7	16.7	33.3	33.3
10. Leather x Leather:*				
ssNn x *ssNn*	0	33.3	0	66.7

* In these crosses 25% of offspring (*NN*) die; the ratios among viable fish are shown.

Figure 2.4.1. Redcap oranda goldfish; this breed has specific red "hood" on the head.

Figure 2.4.2. Long-fin (A) and normal short-fin (B) koi.

popular in the United States. Recently Gomelsky et al. (2011) have reported information on inheritance of long fins in koi. Crossing of two long-fin koi gave Mendelian segregation 3:1 (long-fin : short-fin) in progeny while crossing of long-fin koi with short-fin fish gave Mendelian segregation 1:1. Based on these data it was concluded that, the same as in zebrafish, appearance of long fins in koi is under control of a dominant mutation of one gene (*Lf/lf*). Fish with genotypes *LfLf* and *Lflf* have long fins while fish with genotype *lflf* do not have this trait.

Chapter 3. Inheritance of Quantitative Traits in Fish

3.1. Quantitative Traits: Main Features and Principles of Inheritance

In previous chapter the inheritance of qualitative traits has been described. The second main type of traits is quantitative traits. Quantitative traits are characterized by the following features:

- Have continuous variation;
- Do not have discrete phenotypic classes but are expressed by certain numerical value;
- Are determined by large number of genes;
- Their manifestation depends on environmental conditions.

Very numerous and different traits in fish are quantitative according to their variability and inheritance. Here are some examples of quantitative traits in fish: weight and length, survival, fecundity, meristic (countable) traits (e.g. number of vertebra, rays in fins), physiological traits (e.g. content of fat, hemoglobin).

As mentioned above, quantitative traits are characterized by **continuous variation**. Most quantitative traits have a normal distribution, which is expressed graphically as a smooth curve with a symmetrical location of frequencies decreasing towards both sides from a mean value (bell-shaped curve) (Figure 3.1.1).

Figure 3.1.1. Graph of Normal Distribution.

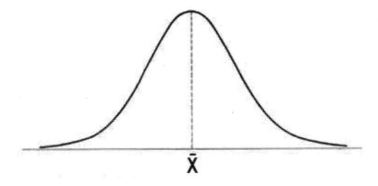

$\bar{X}$

A normal distribution is characterized by the following main indices -

Mean:

$$\overline{X} = \sum X_i / n$$

Standard deviation:

$$\sigma = \sqrt{\sum (X_i - \overline{X})^2 / (n-1)}$$

Standard deviation reflects the rate of trait variability. According to the rules of normal distribution, 99.7% of all variants are placed in the range $\pm 3\sigma$.

The other index - **variance** (σ^2) is usually used for determination of different components of variation for quantitative traits. It is determined as the square of the value of standard deviation, or:

$$\sigma^2 = \sum (X_i - \overline{X})^2 / (n-1)$$

The same principles of gene transmission are relevant for both qualitative and quantitative traits. It was shown for the first time (in the early 1910s) by the Swedish scientist H. Nilsson-Ehle in studies on traits with additive gene action. In Nilsson-Ehle's experiments, wheat with dark-red grain color (genotype *AABB*) was crossed with wheat with white grain color (genotype *aabb*). All F_1 hybrids had intermediate pink grain color. In F_2, five distinct phenotypic classes ranging from white to dark-red with the 1:4:6:4:1 ratio was observed. Nilsson-Ehle suggested that two independently assorting genes were involved in determination of the grain color. Each gene had two alleles with incomplete dominance; one allele for red color, and another allele for white color. Resulting grain color is determined by the number of alleles of red color in the genotype. After crossing the lines, which differed with regard to 3 pairs of alleles (*AABBCC* - dark-red grain, *aabbcc* - white grain), the color ratio in F_2 was 1:6:15:20:15:6:1. This shows that the increase in number of genes controlling the trait results in the increase of the number of phenotypic classes. The difference between classes decreases, and the distribution becomes continuous and similar to normal, which is typical for quantitative traits.

Thus, the quantitative traits are under control of many genes with additive action. Because many genes are involved, inheritance of this type is often called **polygenic**. Each of such genes may have either an **additive allele**, which contributes a set amount to the phenotype, or **nonadditive allele**, which does not contribute quantitatively to the phenotype. The resulting phenotype of a given individual depends on the ratio of additive and non-additive alleles in the genome.

Traits that are inherited by additive gene action (like a wheat color in Nilsson-Ehle's experiments) are described also in fish. Black color in molly (*Poecilia sphenops*) (Figure 3.1.2) is determined according to the same principles. Schröder (1976) has shown that the black color in aquarium strains of molly is determined by two additive genes with two alleles each (*M/m* and *N/n*). The scheme and results of crosses between black (genotype *MMNN*) and non-mottled fish (genotype *mmnn*) are given in Figure 3.1.3. Resulting ratio in F_2 (1:4:6:4:1) is typical for additive gene action when two genes are involved.

Figure 3.1.2. Black molly (*Poecilia sphenops*).

Figure 3.1.3. Inheritance of black color in molly (*Poecilia sphenops*)(additive gene action).

P: ♀ *MMNN* (black) x ♂ *mmnn* (non-mottled)

F₁: *MmNn* (moderately mottled)

F₁ x F₁: ♀ *MmNn* x ♂ *MmNn*

GAMETES

♀ \ ♂	MN	Mn	mN	mn
MN	*MMNN* black	*MMNn* strongly mottled	*MmNN* strongly motled	*MmNn* moderately mottled
Mn	*MMNn* strongly mottled	*MMnn* moderately mottled	*MmNn* moderately mottled	*Mmnn* slightly mottled
mN	*MmNN* strongly mottled	*MmNn* moderately mottled	*mmNN* moderately mottled	*mmNn* slightly mottled
mn	*MmNn* moderately mottled	*Mmnn* slightly mottled	*mmNn* slightly mottled	*mmnn* non-mottled

F₂: 1 (black) : 4 (strongly) : 6 (moderately) : 4 (slightly) : 1 (non-mottled)
mottled   mottled   mottled

3.2. Genetic Analysis of Quantitative Traits. Concept of Heritability.

The continuous character of variation and the absence of definite phenotypic classes make it impossible to apply to quantitative traits the usual methods of genetic analysis that are applied to qualitative traits. Analysis of any quantitative trait is based on the creation of a mathematical model of inheritance. This model describes the character of trait variability, evaluates the possible number of genes controlling the trait, determines which proportion of phenotypic variation has genetic background.

After crossing of two strains, which are different with regard to some quantitative trait, the F_1 has the intermediate value of trait development. The variability observed in F_1 is not so high. In F_2 the mean value remains about the same as in F_1 but the variability is much higher than that observed in F_1. For the first time this mode of inheritance of quantitative traits was described by American geneticists Emerson and East in 1913, when they studied inheritance of ear length in maize. In fish this type of inheritance was observed, for example, in study performed by Wilkens (1971) when he has crossed cave-living blind, having extremely reduced eyes fish (genus *Astyanax*) with related species which inhabited surface water bodies and had a normal development of eyes. In this case the rate of eye reduction was inherited as a typical quantitative trait. The distributions of fish from parental forms (blind and normal fish) as well as F_1 and F_2 hybrids with regard to eye size are given in Figure 3.2.1. You can see in the figure that F_1 hybrids between two forms had intermediate eye size and were relatively uniform. Hybrids F_2 had mean value of eye size close to that of F_1 but were very variable and ranged from values of trait observed in both parental forms.

There are special formulae which allow calculating the approximate number of genes that control the inheritance of a quantitative trait based on the difference in mean values and rates of variability in F_1, F_2 and backcross progenies. By analyzing the results of crosses between blind cave and normal surface fish Wilkens (1988) estimated that about 6-7 genes are involved in determining eye size in these fish.

Phenotypic variability of any quantitative trait is the result of joint action of genes and environmental factors. This can be expressed as:

Phenotype = Genotype + Environment

Variation is usually measured in terms of variance. Therefore the above formula may be written as:

$$\sigma^2_{Ph} = \sigma^2_G + \sigma^2_E$$

where σ^2_{Ph} is the phenotypic variance, σ^2_G is genetic variance, and σ^2_E is environmental variance.

Figure 3.2.1. Distribution of fish from parental forms and their hybrids with regard to eye size; Sa – cave fish with reduced eyes, Ast – fish with normal eyes from surface water bodies, ME – units of measurements; (from Wilkens 1971, Evolution 25:530-544; reproduced with permission from Wiley/Blackwell).

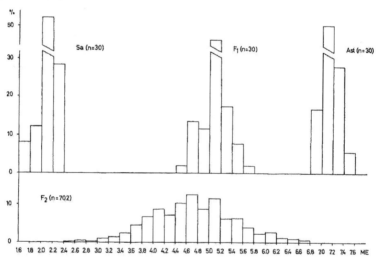

One of the most important tasks in the analysis of quantitative traits is an evaluation of the portion of genetic variability in the observed phenotypic variability. The index measuring this portion is called **heritability** (designated as h^2). It is determined as the ratio of genetic variance to total phenotypic variance:

$$h^2 = \sigma^2_G/\sigma^2_{Ph} \quad \text{or} \quad h^2 = \sigma^2_G/(\sigma^2_G + \sigma^2_E)$$

Heritability measures which part of the total phenotypic variability is caused by genetic factors. It may vary from 0 to 1 (or from 0 to 100%). If h^2 equals 0, it means that the observed phenotypic variability is caused by environmental factors only, and genotypic variation is completely absent. An example of such variability is the rearing of genetically identical progenies (clones) obtained from fish reproducing by means of natural gynogenesis (Chapter 1). (Clones also may be obtained in species reproducing by normal sexual mode by means of induced gynogenesis; Chapter 6).

Genetic variance (σ^2_G) includes both additive (σ^2_A) and non-additive (σ^2_{NA}) components of variability:

$$\sigma^2_G = \sigma^2_A + \sigma^2_{NA}$$

Additive variance (σ^2_A) is caused by genes that act additively towards the development of a given quantitative trait (as in the examples with color of wheat grain or black color in molly). The genes contributing to additive variance have their own effect, which does not depend on other genes. Non-additive variance (σ^2_{NA}) results from gene combinations. Non-additive variance consists of two components: σ^2_D, which is dominance variance, and, σ^2_I which is interactive variance:

$$\sigma^2_{NA} = \sigma^2_D + \sigma^2_I \quad \text{or} \quad \sigma^2_G = \sigma^2_A + \sigma^2_D + \sigma^2_I$$

Dominance variance (σ^2_D) results from interaction of alleles of one gene (dominance, overdominance - advantage of heterozygotes over both types of homozygotes). Interactive variance (σ^2_I) results from interaction of different genes (epistasis, complimentary gene action etc.).

Non-additive genetic variance (σ^2_{NA}), which results from interaction of alleles or genes, cannot be kept by selection. For selection purposes, additive genetic variance is important. Therefore for practical usage in selection the index narrow-sense heritability is used. It shows which part of total phenotypic variation is caused by additive genes:

$$h^2 = \sigma^2_A / \sigma^2_{Ph}$$

3.3. Methods of Heritability Determination

There are three main methods for determination of heritability in fish.

1. Determination of heritability from regression of the trait value between parents and offspring

Determination of heritability by this method is based on the suggestion that if there is some genetic basis to the development of a trait, then offspring should resemble their parents. The change of trait in offspring and parents is described by an equation of linear regression:

$$y = a + bx$$

where y and x are the mean values of the trait in offspring and the parents, respectively; b is the regression coefficient, and a is constant. This method is based on the calculation of regression coefficient b, which reflects the rate of trait variation in offspring when it is changed in parents per unit. When the offspring are compared with both parents, heritability is equal to the regression coefficient:

$$h^2 = b$$

If the offspring are compared with only one parent the formula is transformed to:

$$h^2 = 2b$$

This method was used mostly for determination of heritability for some meristic (countable) traits in fish. For example, Kirpichnikov (1981) has obtained the following regression equation for the number of soft rays in the dorsal fin in common carp: $y = 7.32 + 0.61x$. Based on these data heritability (h^2) was equal b or 0.61.

2. Determination of heritability from effectiveness of selection

The effectiveness of selection per one generation may be determined according to the formula:

$$R = h^2 S$$

where R is the change of the selected trait (response of selection), S is selection differential, i.e. the difference in the mean value of the trait in selected individuals and the mean value in the whole population before selection, and h^2 is heritability. (Materials on fish selection are given in details in Chapter 5). According to the formula given above the value of heritability may be determined as:

$$h^2 = R/S$$

Using this formula it is possible to calculate the value of heritability according to results obtained by selection. Realized heritability shows which fraction of the difference between selected individuals and the whole population of the parent generation is retained in the offspring. For example, Dunham and Smitherman (1983) calculated realized heritability for body weight for channel catfish. Mean response of selection for three strains of channel catfish (R) was 64 g while selection differential was 190 g; the calculated value of heritability was 0.34 (64/190).

3. Determination of heritability by partitioning of total phenotypic variance into its components

This method of heritability determination has several consecutive steps:

1. Production of a large number of related progenies according to a specific scheme;
2. Collecting data on some quantitative trait in progenies;
3. Variance analysis of obtained data for partitioning (splitting) of observed phenotypic variability into genetic and environmental components.

Related progenies for determination of heritability are usually obtained according to so-called hierarchic complexes, which can be of two possible types. In hierarchic complexes of the first type each of several females are crossed with several males; in the opposite scheme each male is crossed with several females. Fish originating from the same parents are called full-sibs, while fish having only one common parent are called half-sibs. Obtained progenies are raised separately with several replicates under similar conditions. The data obtained are treated by means of analysis of variance (ANOVA). Application of ANOVA for the partitioning of general phenotypic variance to its genetic and environmental components is based on the assumption that variance between replicates of the same cross is caused by environmental factors while among-family variance between crosses reflects genetic differences. (The detailed description of this method can be found in textbooks on quantitative genetics).

Fish characters such as external fertilization and high fecundity make it possible to produce large number of progenies by hierarchic complexes. For example, Reagan et al. (1976) crossed each of 17 catfish males with 2 different females; 34 progenies were produced and analyzed. Crandell and Gall (1993) crossed each of 18 rainbow trout females with 3 different males; 54 progenies were produced and analyzed.

Numerous studies on heritability determination in fish, performed in different species, have shown that among all quantitative traits the meristic (countable) traits (such as the number of vertebrae, rays in fins etc.) have the highest level of heritability (0.6-0.9). The values of these traits are determined at early stages of ontogeny and later they do not depend on environmental conditions. An average level of heritability (0.3-0.6) is typical for some morphological, reproductive and physiological traits. Most important traits of fish productivity, such as weight and survival, have relatively low level of heritability (usually 0.1-0.4). These quantitative traits are most dependent on environmental conditions. Besides this, heritability may decrease as a result of previous selection.

Development of DNA technology opened new possibilities for genetic analysis of quantitative traits. Application of microsatellites markers makes possible parentage determination (see Chapter 4). It gives opportunity to raise fish from different families communally in one pond or tank from very early stages. Communal rearing minimizes environmental component of variance. In last 5 years several studies on determination of heritability using microsatellite markers have been performed in several aquaculture species (e.g. common carp, sea bass and red drum). Also, locations of genes controlling quantitative traits in chromosomes may be determined by segregation analysis of associated DNA markers. These locations are called "quantitative trait loci" or QTL (see Chapter 4).

Chapter 4. DNA Genetic Markers and Their Application

4.1. Genetic Markers: Definition and Basic Types

The variability of fish with regard to qualitative morphological traits is very low. Besides rare mutations concerning body color, scale cover or fin length, fish are "very uniform" according to external appearance. Therefore it is very important to find other traits, which could be used as genetic markers.

A **genetic marker** is any characteristic whose inheritance can be followed and which can be used for identification of genetic difference between individuals or groups of individuals. A necessary condition for a genetic marker is the occurrence of **polymorphism**, i.e. observed variability of individuals with regard to this characteristic.

There are two main types of genetic markers:

1. Phenotypic traits, which are controlled by functional genes. This type includes qualitative morphological traits as well as biochemical traits such as differences in protein structure (protein polymorphism) or erythrocyte antigenes (blood groups).
2. Regions of DNA with or without coding function. Genetic markers of this type are called molecular or DNA markers. Molecular markers are based on the difference not in gene products but on the structure of the DNA molecule itself.

Blood groups were the first type of biochemical genetic markers used in fish population genetics. The first data on blood groups in fish were obtained in late 1950s. This direction had been intensively developed up to the discovery of protein polymorphism.

In the middle of the 1960s the method of electrophoresis was elaborated. This technique permitted the separation of protein molecules with regard to their motility when exposed to an electric field. It was shown that the occurrence of several forms (alloforms, or for enzymes - allozymes) is a fairly frequent phenomenon for many proteins. Protein variability, called **protein polymorphism**, is inherited according to Mendelian principles. Protein polymorphism is based on the mutations of genes controlling protein synthesis. Changes in the nucleotide sequence of a DNA molecule, caused by mutation, are reflected in the primary structure of a protein molecule as substitution or falling out of one amino acid. As a result, the electric charge of a protein molecule and its motility in an electric field may change. The resulting difference between protein forms may be detected by means of electrophoresis. For protein polymorphism, codominance is typical, i.e. each allele in heterozygotes is functioning independently, synthesizing its own product. For fish, as well as for other animals, hereditary polymorphism of proteins is a common phenomenon.

In the 1970s the method of electrophoretic separation of proteins made a real revolution in fish population genetics. As indicated in Chapter 2, the variability of fish with regard to qualitative morphological traits is very limited. Therefore, for a long time some morphometric and meristic traits were used for the comparison of fish from different populations. As shown in Chapter 3, these traits are quantitative according to their inheritance, and their manifestation depends on both genotype and environmental conditions. Therefore, the application of these traits for comparison of populations or other groups of fish was restricted. On the contrary, protein polymorphism has discrete inheritance, similar to that of qualitative traits.

Until wide application of DNA markers in the 1990s, protein polymorphism was the most effective tool of fish population genetics. Currently protein polymorphism is rarely used in fish genetics while DNA markers became the main effective tool.

4.2. DNA Markers and Methods of Molecular Genetics

Molecular markers are based on DNA technology methods. A detailed description of molecular genetics methods can be found in textbooks on general and molecular genetics. Basic techniques of molecular genetics, necessary for understanding the learning materials presented in this chapter, are listed and briefly described below:

1. Restriction enzymes cleave ("restrict") DNA at specific nucleotide sequence recognition sites and generate DNA fragments that differ in size. Each restriction enzyme recognizes and cuts DNA only at particular sequence of nucleotides.

2. Southern blot hybridization - This technique, which was developed by Edward Southern in 1975, is used for identifying fragments of DNA that differ in size. First, DNA fragments are cleaved with restriction enzymes, separated on an agarose gel, and then transferred from the gel by blotting onto a nitrocellulose or nylon filter. The filter is hybridized then with a specific radioactive probe. (DNA probes are capable of recognizing a specific DNA sequence in a mixture of many different DNA sequences. A probe is labeled with radioactivity or a chemical that can be detected.) Hybridization allows identification of fragments, which carry sequences complimentary to those on the probe, and determination of their size based on position on an electrophoregram. The specific region appears as a band on an X-ray film exposed to the filter.

3. Recombinant DNA construction - Recombinant DNA is DNA that has been created artificially. DNA from two or more sources is incorporated into a single recombinant molecule. To produce recombinant DNA, a particular fragment of DNA is inserted into a vector, i.e. a DNA molecule that is capable of carrying foreign genetic information. Usually DNA from bacterial plasmids or bacteriophages is used as vectors. DNA fragments are generated by using of restriction enzymes while enzyme DNA ligase creates bonds between fragments. The resulting recombinant plasmid or phage DNA is then introduced into bacterium, where it can replicate. A collection of cloned DNA fragments is called DNA library.

4. Polymerase chain reaction (PCR) was devised by Kary Mullis in 1986. PCR involves replicating target regions of DNA, which are flanked by short regions of known sequences. The oligonucleotide primers, which are complementary to each of the flanking regions, are needed to begin synthesis of the target region of DNA.

These are combined with a small sample of genomic (template) DNA, free nucleotides, a reaction buffer, and a heat-stable form of DNA polymerase. During a series of heating and cooling cycles, the DNA is denatured into single-stranded molecules, the two primers form base pairs with their complementary sequences in the single-stranded DNA at the boundaries of target region, and then DNA polymerase replicates the region downstream from each primer. The amount of target DNA doubles with each cycle. The resulting amplified DNA can then be separated by gel or capillary electrophoresis.

5. DNA sequencing has become a routine procedure since the development of the dideoxy chain termination method by Frederick Sanger in 1975. This technique relies on using modified nucleotides, which terminates the synthesis reactions at a specific nitrogenous base (A, G, C, or T). The resulting DNA fragments are then visualized on a polyacrylamide gel. Sanger sequencing is the basis for automated sequencing techniques.

Molecular DNA-based genetic markers are widely used now in human genetics, forensics, and medicine. They have also become popular in applied genetics of animals including fish genetics. In further parts of this chapter (4.3 and 4.4) the description of the main DNA markers used in fish genetics and directions for their application are presented.

4.3. Types of DNA Markers

4.3.1. Restriction Fragment Length Polymorphism of Nuclear and Mitochondrial DNA

Restriction Fragment Length Polymorphism (RFLP, pronounced 'riflip') was the first elaborated type of DNA marker. RFLP appeared when fragments of DNA, cleaved by specific restriction enzymes, were separated by gel electrophoresis and detected by subsequent Southern blot hybridization. Differences in the length of the generated fragments resulted from mutations or changes in the base sequence of DNA.

Due to limited variability and some technical difficulties, RFLP of nuclear (chromosomal) DNA was not used as a practical genetic marker in fish. However, this technique was successfully applied for the analysis of mitochondrial DNA (mtDNA). Mitochondrial DNA in animal cells is a relatively small circular molecule. It consists of 16,000-26,000 base pairs (or 16-26 kb - kilobase pairs) in length. The mitochondrial genome is haploid, maternally inherited and non-recombining. Because of the small size of the mitochondrial DNA, RFLP was effective for its analysis. It was possible to visualize fragments of mitochondrial DNA, which resulted from the action of restriction enzymes, directly (without Southern blot hybridization) by staining following electrophoresis of the DNA in an agarose gel.

The first studies on variability of mitochondrial DNA in fish were published in the middle of the 1980s. As an example, the profiles of mtDNA isolated from the cultured stock of gilthead sea bream (*Sparus aurata*) are shown in Figure 4.3.1. You can see that two different haplotypes (designated as "a" and "b") were identified after treatment of mtDNA with any of 4 different restriction enzymes (Xho I, Ava I, Bgl I, and Pvu II).

Mitochondrial DNA polymorphism analysis is still actively used in fish genetics. The RFLP technique was further developed by application of PCR for amplifying mtDNA fragments (so called PCR-RFLP technique). Complete sequencing of mtDNA from several fish species was also performed. Polymorphism of mtDNA may be used for comparison of fish from different populations or species and phylogenetic studies. Intrapopulation variability (i.e. variation of fish inside a population) of mtDNA is usually not high. Maternal inheritance and the absence of recombination restrict the application of mtDNA polymorphism.

4.3.2. Multilocus Minisatellites

DNA of eukaryotic organisms, along with single-copy functional genes, contains large amounts of repetitive sequences. There are several types of repetitive

Figure 4.3.1. Profiles of mitochondrial DNA isolated from the population of gilthead sea bream (*Sparus aurata*) maricultured in Eilat, Israel; "a" and "b" represent two different patterns obtained with the same enzyme (from Funkenstein et al. 1990, Aquaculture 89:217-223; Copyright 1990, reproduced with permission from Elsevier).

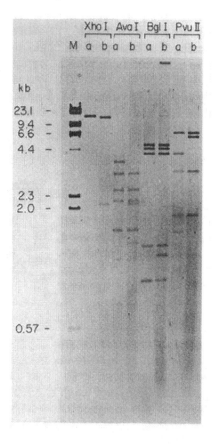

DNA. One type of repetitive DNA is known as **minisatellites**. These regions of DNA do not have a coding function and are formed from head-to-tail repetitions of certain short sequences. A remarkable feature of minisatellites is that the number of copies of the repeating DNA sequence may vary. This results in variation in the size of these DNA regions between individuals and permits them to be used as molecular markers.

The size of repeating DNA sequence in minisatellites varies from 10 to 50 base pairs, while the number of repeats ranges from 2 to more than 100. As was mentioned above, minisatellites are variable due to the variation in the number of repeat units. Variability of minisatellites with regard to the number of repeat units and the possibility of using this variability for individual genetic identification was devised by Alec Jeffreys in 1985. This technique was called **multilocus DNA fingerprinting**.

Variability for minisatellite loci are detected by cutting genomic DNA with restriction enzymes, the separation of resulting fragments according to their size by gel electrophoresis, Southern blotting to nylon, and hybridizing to repeat sequence probes. When visualized on a Southern blot, the DNA fragments appear as a series of bands. This pattern reflects total variability of a given individual for many minisatellite loci. Because of the high level of polymorphism of minisatellites, each individual has unique combinations of alleles and, therefore, a unique pattern of bands. The patterns of bands may be completely identical only in identical twins in humans or in genetically identical fish from clones.

The development and application of multilocus DNA fingerprinting technique was a real revolution in human genetics, namely in forensics and paternity testing. Along with application in human genetics, multilocus DNA fingerprinting technique was used in animal genetics. The first articles reporting the application of this method for fish were published in the end of the 1980s.

As an example, Figure 4.3.2 represents multilocus DNA fingerprints of rainbow trout (from study performed by Palti et al. 1997). You can see in the figure that each individual fish has specific band combinations. For comparison of DNA fingerprints between individual fish or groups of fish a special band-sharing index was calculated.

In fish, multilocus DNA fingerprinting was especially valuable for the confirmation of the genetic identity of clones obtained by means of induced

Figure 4.3.2. DNA fingerprints of 17 rainbow trout individuals from one stock (Oregon State University). The two flanking lanes and the tenth lane from left are molecular weight standards; molecular weights are given in base pairs (from Palti et al. 1997, Aquaculture 149:47-56; Copyright 1997, reproduced with permission from Elsevier).

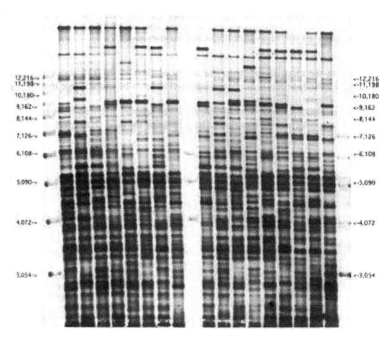

gynogenesis. Figure 4.3.3 presents multilocus DNA fingerprints of normal diploids of red sea bream, fish from homo- and heterozygous clones and the mother of these clones (M). (Methods of obtaining homo- and heterozygous clones in fish using induced gynogenesis are described in Chapter 6). You can see in the figure that normal fish are heterogeneous while fish in both clones are completely uniform. Note that fingerprints of fish from the homozygous clone are identical to the fingerprint of mother from which this clone originated; this identity was expected taking into account the method of homozygous clone production (see Chapter 6).

Figure 4.3.3. Multilocus DNA fingerprints of normal diploids (Normal-2N) of red sea bream (*Pagrus major*), fish from heterozygous (Hetero-clone) and homozygous (Homo-clone) clones and the mother of these clones (M) (from Kato et al. 2002, Aquaculture 207:19-27; Copyright 2002, reproduced with permission from Elsevier).

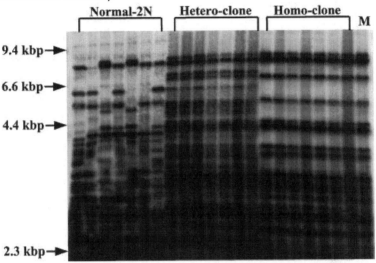

As was mentioned above, the bands visible in multilocus DNA profiles could not be assigned to certain loci. This made it impossible to calculate the allele frequencies in populations and, therefore, restricted the application of multilocus DNA fingerprinting for genetic population analysis. A further step was the development of single-locus minisatellite probes. In this case DNA profiles have contained alleles only of one locus. However, single-locus minisatellites did not become a popular type of DNA marker in fish genetics due to development of PCR-based microsatellites, which have similar mode of inheritance. (PCR is a much simpler technique). Currently minisatellites are not practically used in fish genetics.

4.3.3. Microsatellites or Simple Sequence Repeats

Microsatellites represent another type of repetitive DNA. Microsatellites, also called Simple Sequence Repeats (SSRs), are composed of tandemly repeated

2-6-base units (motifs). Similarly to minisatellites, the allelic variability of microsatellites is based on the difference in the length of fragments caused by the different number of repeated units. The same as minisatellites, microsatellites do not have any coding function. The number of repeats in microsatellite loci usually varies from 8 to 50. Because of their relatively small sizes, microsatellite regions may be amplified using PCR. The isolation of microsatellite loci from a genomic library for a new species or group of species is the first step for the investigation of microsatellite polymorphism. The sequence information for the flanking DNA is needed to allow the synthesis of specific PCR primers. Following primer design, products can be amplified and separated through polyacrylamide gel or capillary electrophoresis.

Microsatellites are locus-specific and codominant markers. Microsatellites are also highly variable, and the number of alleles for one locus may be more than 30. Heterozygotes for microsatellite loci have two bands on a gel which represent two alleles while homozygotes have one band (one allele). Microsatellite alleles are designated by numbers which indicate the size of corresponding PCR product in base pairs (bp). As an example, the Figure 4.3.4 (from study performed by Taniguchi et al. 1999) represents variation of 14 red sea bream individuals for locus *Pma 4*. Each lane on polyacrylamide gel represents genotype of one individual fish. Alleles were sized relative to a standard M13 sequence ladder shown on the right side of the figure. The numbers under each lane show the genotype of the corresponding individual fish. For example, the DNA profile at the extreme left side shows a fish with genotype 91/111, i.e. this fish is heterozygous with the size of one allele 91 bp and the second allele 111 bp. You can see on the gel two homozygous fish (125/125, 115/115) whose profiles have only one band. (Note that each main allele band is accompanied with several so called stutter bands).

Currently DNA sequencers (Genetic Analyzers) are frequently used for genotyping microsatellite alleles. In this case PCR products are separated via electrophoresis inside very thin tubes or capillaries. The capillaries are screened for laser-induced fluorescence intensity. In the resulting computer-generated graphs

Figure 4.3.4. Genotype variability of red sea bream for microsatellite locus Pma4 (from Taniguchi et al. 2009; pp. 206-218 in: Genetics in sustainable fisheries management; reproduced with permission from Wiley/Blackwell).

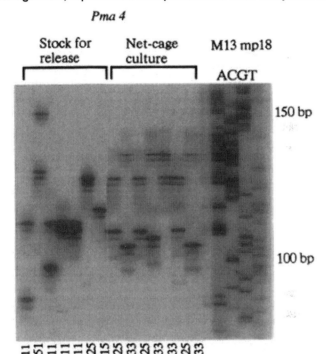

Pma 4

(electrophoregrams) the peaks of fluorescence intensity correspond to the size of allele fragments. Figure 4.3.5 presents graphs of fluorescence intensity for microsatellite genotyping of freshwater prawns. You can see in Figure 4.3.5.A that based on location of fluorescence intensity peaks the analyzed individual has genotype 263/277. Figure 4.3.5.B presents graph for homozygous individual with genotype 298/298; homozygotes have only one fluorescence intensity peak. (Note that in both graphs the highest allele peak is preceded by several smaller, so-called stutter peaks.)

Figure 4.3.5 Graphs of fluorescence intensity for genotyping of freshwater prawn for one microsatellite locus (*Mbr-1*) (courtesy of Kyle Schneider).

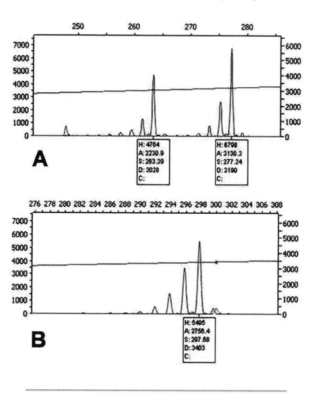

Numerous experiments conducted in fish (as well as in other animals) showed that microsatellites are inherited according to Mendel's principles. For example, Alsaqufi (2011) has investigated inheritance of microsatellite markers in koi. Five progenies (families) were obtained by individual crossing of females and males; fish parents and offspring have been genotyped for 10 microsatellite loci and observed segregations in progenies have been compared with theoretical ratios. Below, as an example, the theoretical segregations in some crosses for two microsatellite loci (*Cca02* and *MFW26*) are shown:

Locus *Cca02*

♀175/175 x ♂177/177 → 100% 175/177

♀157/177 x ♂175/175 → 50% 157/175 : 50% 175/177

Locus *MFW26*

♀128/144 x ♂ 128/144 → 25% 128/128 : 50% 128/144 : 25% 144/144

♀128/144 x ♂ 144/154 → 25% 128/144 : 25% 144/144 : 25% 128/154 : 25%144/154

♀140/154 x ♂ 128/144 → 25% 128/140 : 25% 128/154 : 25% 140/144 : 25%144/154

You can see that the genotypic ratio in progeny is determined by the simple combination of parental alleles. Depending on the genotypes of parents, the expected ratios in progenies are 1:0, 1:1, 1:2:1, or 1:1:1:1.

The first studies on application of microsatellites as genetic markers in fish were published in the beginning of the 1990s. From this time, microsatellites were identified and analyzed in many important fisheries and aquaculture species. Because of their very high variability, discrete character of inheritance, codominance, and established standard techniques for their analysis, microsatellites are regarded now as the most popular and powerful type of DNA markers in fish genetics. As you will see in the next part of this chapter (4.4), the polymorphism for microsatellite loci may be used for many applied tasks in fish genetics.

4.3.4. Random Amplified Polymorphic DNA (RAPD)

Random Amplified Polymorphic DNA (or RAPD, pronounced "rapid") markers are produced by PCR (the same as microsatellites). However, in this case amplification is initiated by application of non-specific randomly chosen short primers. Currently hundreds of different RAPD primers are commercially available. The ability of primers with random sequence to induce amplification is based on the fact that a genomic DNA molecule is very large and by chance there can be short regions which are complimentary to the tested primers. Primers are used individually, not in combination with a second primer as would be the case for standard PCR. Because of this, amplified fragments are those regions of the genome that are flanked by "inward-oriented" sequences complimentary to the primer. Allelic variation is based on the presence (+) or absence (-) of particular amplification products, which are separated by gel electrophoresis.

In contrast to microsatellites, which demonstrate codominance, RAPDs are dominant, i.e. the presence of a RAPD band does not allow distinction between homozygotes (+/+) and heterozygotes (+/-). RAPD markers show typical dominant Mendelian inheritance in crosses.

Novelo (2008) has investigated inheritance of RAPD markers in common carp. In Figure 4.3.6 the RAPD profiles of parent female (F), parent male (M) (two left lanes) and 32 fish from F_1 progeny obtained by their cross are presented. We will consider inheritance of two strongly amplified RAPD markers (754 and 803 bp), which can be regarded as two different loci (see Figure 4.3.6). You can see that at locus 754 bp the male (M) and all F_1 individuals had an amplification product while the female (F) did not have this band. This shows that the male was homozygous for the dominant allele (genotype +/+), the female was homozygous for the recessive allele (-/-) and all F_1 fish were heterozygotes +/-. At the other locus (803 bp) the female was heterozygous (+/-) and its crossing with the male -/- resulted in segregation 1 with band (+/-) : 1 without band (-/-) in F_1 progeny (L is a standard ladder used for determination of fragment size).

The important benefit of the RAPD technique is that it does not require any previous DNA sequence data. However, these markers have also some negative features. Inability to distinguish between homozygotes and heterozygotes makes RAPD markers less useful for genetic analysis of populations. Also, it is regarded that RAPD markers have questionable reproducibility.

4.3.5. Amplified Fragment Length Polymorphism (AFLP)

The technique to detect **Amplified Fragment Length Polymorphism** (AFLP) combines the features of RAPD and RFLP. As in RAPDs, AFLP is based on PCR and does not require any known sequence information of the target genome.

The AFLP technique is based on selective amplification of digested genomic DNA by a series of extended primers. AFLPs are DNA fragments (80-500 bp)

Figure 4.3.6. RAPD profiles of common carp parents (F and M) and offspring (F1) obtained by using primer UBC109 (from Novelo 2008; reproduced with permission).

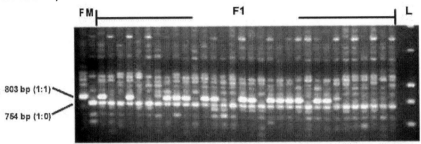

803 bp (1:1)

754 bp (1:0)

obtained by digestion of genomic DNA with two restriction enzymes, a frequent and a rare cutter, followed by ligation of oligonucleotide adaptors to the fragments and selective amplification by PCR. The AFLP-technique simultaneously generates fragments from many genomic sites that are separated by electrophoresis. Like RAPDs, AFLPs are also dominant markers.

Figure 4.3.7 presents AFLP profiles for 10 channel catfish strains with the B2 primer combination. Strains are indicated at the top by letters; three individuals were analyzed from each strain. Arrows on the right indicate bands shared by all 10 strains. Among the 78 amplified bands shown in Figure 4.3.7, 53 demonstrated various levels of polymorphism.

Due to their high genomic abundance and random distribution throughout the genome, AFLP markers are used for gene mapping.

4.3.6. Single Nucleotide Polymorphism (SNP)

Single Nucleotide Polymorphism markers (or SNPs; pronounced "snips") are single base changes in a DNA sequence. In theory each SNP locus can have 4 alleles, which correspond to 4 different bases: adenine (A), cytosine (C), thymine (T) or guanine (G). In reality, however, each SNP locus has only two alleles. The exchanges of either two purine bases (A and G) or two pyrimidine bases (C and T) are observed more frequently. SNPs are codominant markers.

Figure 4.3.7. AFLP profiles for 10 channel catfish strains (from Liu et al. 1999; reproduced with permission from American Fisheries Society).

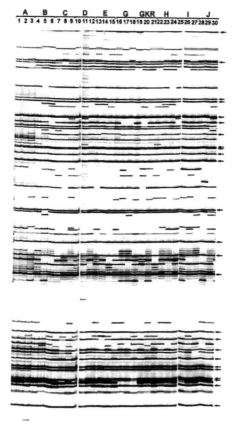

Single Nucleotide Polymorphism cannot be regarded as a new type of DNA marker. It was described in the 1970s when DNA sequencing was started. However, popularity of SNPs increased drastically in recent years. One of the main reasons of this increase is that methods for detection of SNPs can be automated using modern methods of DNA technology allowing to identify thousands of SNPs. The other advantages of SNPs are their extreme abundance and even distribution in the genome. Currently Single Nucleotide Polymorphism is regarded as a very effective tool in genomics (Chapter 7), the part of genetics devoted to the study of structure and function of entire genomes. In contrast to microsatellites, which usually are located in non-coding regions of DNA, many SNPs are located in protein coding regions, i.e. inside functioning genes. Therefore it is possible to use SNPs to track associations with different traits, for example, growth rate or resistance to disease.

Currently SNP markers are not widely used in applied fish genetics (as compared, for example, with microsatellites) but their role increases.

Smith et al. (2005) studied single nucleotide polymorphism in 16 North American and Asian populations of Chinook salmon. A total of 10 SNP loci have been genotyped. The data on allele frequencies in 7 populations are presented in Table 4.3.1. You can see in the table that each SNP locus has 2 alleles, which are different bases (A, C, G or T). Many populations (abbreviated names of populations are given in the upper line with numbers of fish in samples in parentheses) differ greatly with regard to allele frequencies.

4.4. Application of DNA-based Genetic Markers in Aquaculture and Fisheries

Directions of DNA markers application may be divided into two main groups. The first group includes the following "classical" directions, for which protein polymorphism was used as the main tool prior to elaboration of DNA markers:

- Evaluation of within-population and between-population genetic variability
- Monitoring the level of genetic variability
- Identification of species and distant hybrids

DNA-based genetic markers are more powerful tools compared with protein polymorphism. Application of molecular markers provides opportunities to solve novel, more complex tasks. The second group includes these new directions of DNA markers application:

- Parentage determination
- Genetic mapping
- Identification of Quantitative Trait Loci (QTL)

Table 4.3.1. Observed allele frequencies for 10 SNP loci in 7 populations of Chinook salmon); allele names correspond to the four standard nucleotide bases (A, G, C, T) (from Smith et al. 2005, reproduced with permission from American Fisheries Society).

Locus	Allele	Bist (94)	Ston (95)	Togi (91)	Nush (95)	Ayak (93)	Moos (46)	Kena (92)
Ots_GH2	A	0.937	0.411	0.843	0.789	0.785	0.907	0.783
	T	0.063	0.589	0.157	0.211	0.215	0.093	0.217
	F_{IS}	−0.062	−0.169	−0.096	−0.071	0.007	−0.091	−0.14
Ots_Prl2	A	0.659	0.394	0.549	0.668	0.337	0.772	0.528
	G	0.341	0.606	0.451	0.332	0.663	0.228	0.472
	F_{IS}	−0.122	−0.088	−0.077	0.147	−0.016	−0.162	−0.042
Ots_Tnsl	A	0.017	0.106	0.102	0.129	0.081	0.045	0.056
	G	0.983	0.894	0.698	0.871	0.919	0.955	0.944
	F_{IS}	−0.011	−0.002	−0.108	−0.143	0.063	−0.032	−0.053
Ots_Ots2	A	0.112	0.037	0.12	0.101	0.02	0.028	0.101
	G	0.888	0.963	0.88	0.899	0.98	0.974	0.839
	F_{IS}	0.056	−0.033	−0.011	−0.107	−0.014	−0.013	0.069
Ots_MHC1	A	0.594	0.632	0.606	0.536	0.287	0.197	0.424
	G	0.406	0.368	0.394	0.464	0.713	0.803	0.576
	F_{IS}	0.265	0.01	0.006	0.002	−0.066	0.1	0.071
Ots_P53	A	0.426	0.271	0.38	0.489	0.563	0.463	0.309
	G	0.574	0.729	0.62	0.511	0.437	0.538	0.691
	F_{IS}	0.26	−0.098	0.008	0.184	0.025	0.258	0.085
Ots_MHC2	G	0.973	0.165	0.114	0.126	0.075	0	0.029
	T	0.027	0.835	0.886	0.874	0.925	1	0.971
	F_{IS}	−0.022	0.04	−0.004	−0.004	−0.076	NA	0.387
Ots_P450	A	0.128	0.234	0.188	0.183	0.192	0.104	0.319
	T	0.872	0.766	0.812	0.617	0.808	0.896	0.681
	F_{IS}	−0.141	−0.063	−0.225	0.142	−0.082	0.349	0.094
Ots_SL	A	0.781	0.853	0.59	0.697	0.434	0.458	0.819
	G	0.219	0.147	0.41	0.303	0.566	0.542	0.181
	F_{IS}	0.053	0.084	−0.213	−0.111	0.203	−0.049	0.03
Ots_C3N3	G	1	1	1	1	1	1	1
	T	0	0	0	0	0	0	0

- Marker-assisted selection
- Identification of genetic sex

In this part of Chapter 4 the main directions of DNA markers application are briefly described.

4.4.1. Evaluation of Within-Population and Between-Population Genetic Variability

Data on DNA markers gives the possibility to evaluate genetic variability both within and between groups of fish.

The allele frequencies and levels of heterozygosity observed in populations are compared with estimates calculated according to Hardy-Weinberg equilibrium.

Hardy-Weinberg law is one of the basic principles of population genetics. It states that in any sufficiently large, randomly mating population, where the influence of selection and inbreeding is insignificant, the following equilibrium ratio of genotypes is established:

$$p^2\text{AA} : 2pq\text{AB} : q^2\text{BB}$$

where p and q are frequencies of alleles A and B, respectively; AA, BB, and AB are corresponding genotypes. The Hardy-Weinberg law was named after G. Hardy and W. Weinberg, who derived it independently of each other in 1908. The Hardy-Weinberg law means that the process of sexual reproduction does not by itself (without selection or other factors) change the overall composition of genotypes in population. Dominant traits do not automatically increase their frequencies from one generation to the next, and recessive traits do not decrease their frequencies or disappear from populations. Under the conditions listed above, the Hardy-Weinberg equilibrium ratio of genotypes remains constant from generation to generation.

Possible large differences between observed and expected (according to Hardy-Weinberg equilibrium) segregations may be explained by selection, inbreeding, non-random mating or mixing of fish from different populations in sample or by some other factors. There are special statistical tests and computer programs which allow one to evaluate different factors which could shift genotypic ratios in population with excess or deficiency of heterozygotes.

According to the differences in allele composition and frequencies, it is possible to reveal the difference between groups of fish (populations, strains etc.) and to evaluate the rate of this difference. Indices of genetic distances (or similar indices) between groups of fish are calculated and based on these data dendrograms are constructed.

As an example, we will consider a study on application of microsatellite DNA markers for investigation of genetic diversity of cultured and wild populations of the freshwater prawn (*Macrobrachium rosenbergii*) (Schneider 2010). Variability at 5 microsatellite loci has been investigated in two wild and seven cultured populations of freshwater prawn. Genetic characteristics of investigated populations are presented in Table 4.4.1.

Table 4.4.1. Genetic characteristics of wild and cultured populations of freshwater prawn based on microsatellite loci variability (from Schneider 2010; reproduced with permission).

Population	Hawaii 1	Hawaii 2	India Cultured	India Wild	Israel	Kentucky	Mississippi	Myanmar	Texas
N	49	60	45	50	50	60	49	60	50
A	10.80	11.20	18.60	5.40	4.00	7.00	6.60	22.20	6.00
H_o	0.803	0.835	0.872	0.707	0.709	0.657	0.699	0.919	0.736
H_e	0.843	0.851	0.920	0.726	0.579	0.736	0.749	0.936	0.689
F_{is}	0.048	0.019	0.053	0.025	-0.225	0.108	0.067	0.018	-0.070

The following indices have been recorded for every population (see Table 4.4.1):

N - sample size (number of investigated individuals);

A - average number of alleles per microsatellite locus;

H_o - average observed heterozygosity for microsatellite loci;

H_e - average expected heterozygosity for microsatellite loci (based on Hardy-Weinberg equilibrium);

F_{is} - inter-individual fixation index; this index evaluates deficiency or excess of heterozygotes in populations and is calculated based on formula $(H_e - H_o)/H_e$. If F_{is} is positive there is a deficiency of heterozygotes and if negative there is excess of heterozygotes. F_{is} is frequently called 'coefficient of inbreeding'. (You can see also the values of F_{is} in Table 4.3.1 in population study of Chinook salmon using SNPs.)

Based on average number of alleles per locus, the Myanmar and India cultured populations had highest within-population genetic variability while Israel and India wild had smallest genetic variability (see table 4.4.1).

For determination of between-population variability the subpopulation fixation index (F_{st}) and index of genetic distance have been determined. Table 4.4.2 presents matrix of pair-wise comparisons of F_{st} for nine populations of freshwater prawns. The value F_{st} can vary from 0 to 1.0. If F_{st} is equal 0 it means that there is no genetic difference between two populations; if F_{st} is equal 1.0 it means that two populations are completely different. You can see in Table 4.4.2 that two Hawaii population demonstrated the greatest level of similarity $(F_{st}=0.0026)$ while the highest level of genetic difference was recorded between Israel and Texas populations (F_{st}=0.35).

Table 4.4.2. Matrix of pair-wise comparisons of F_{st} for nine populations of freshwater prawns based on microsatellite loci variability (from Schneider 2010; reproduced with permission).

	HI 1	HI 2	IN C	IN W	IS	KY	MS	MY	TX
HI 1									
HI 2	0.0026								
IN C	0.0610	0.0547							
IN W	0.1808	0.1724	0.1419						
IS	0.2623	0.2593	0.2029	0.3092					
KY	0.1570	0.1484	0.1401	0.2163	0.3169				
MS	0.1618	0.1513	0.1320	0.2256	0.3226	0.0443			
MY	0.0567	0.0502	0.0126	0.1112	0.1902	0.1189	0.1126		
TX	0.1854	0.1769	0.1695	0.2449	0.3520	0.0169	0.0534	0.1419	

Based on calculated values of genetic distance a dendrogram (genetic tree) was constructed (Figure 4.4.1). As you can see in the figure, there are two population clusters. The first cluster includes the two Hawaii populations, and the Myanmar and India-cultured population. The second cluster consisted of three U.S. continental populations (Kentucky, Mississippi and Texas). The location of the India-wild and Israel populations illustrates their dissimilarity to the others.

For statistical treatment of experimental data on DNA markers polymorphism in population studies and calculation of different parameters, computer programs are used. There are many computer programs for population genetic analysis whose software can be downloaded from the Internet. Some of them are listed below:

- **GENEPOP** (http://genepop.curtin.edu.au/)
- **TFPGA** (http://www.marksgeneticsoftware.net/tfpga.htm)
- **FSTAT** (http://www2.unil.ch/popgen/softwares/fstat.htm)
- **ARLEQUIN** (http://anthro.unige.ch/software/arlequin/)

Figure 4.4.1. Dendrogram for nine populations of freshwater prawns based on microsatellite loci variability (from Schneider 2010; reproduced with permission).

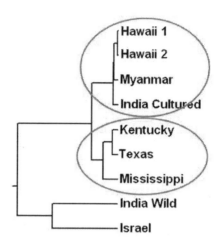

4.4.2. Monitoring the Level of Genetic Variability in Cultured Fish Stocks

The data on DNA markers give possibility to monitor the level of genetic variability in groups of fish. The mean heterozygosity and the number alleles per locus are the main indices for this monitoring. Based on observed decrease in mean heterozygosity and number of alleles it is possible to suggest possible inbreeding and other processes which could diminish genetic variability.

For example, Taniguchi et al. (1999) have found that the mean heterozygosity for microsatellite loci in seventh generation of selectively bred stock of red sea bream was 68% as compared with 87% in natural population. The decrease in the number of alleles per microsatellite locus was even more drastic; this index diminished from 19.8 in natural population down to 7.4 in selected stock. On the basis of these data, the authors (Taniguchi et al. 1999) suggested the occurrence of a bottleneck effect in selective breeding stocks, which resulted from a small number of parents contributing gametes to the next generations.

4.4.3. Identification of Species and Interspecies Hybrids

DNA markers may be used for identification of species and interspecies hybrids. The basic principle of this analysis is determination of species-specific profiles (alleles). RAPD markers are frequently used for this purpose. For example, Barman et al. (2003) investigated RAPD profiles for four species of Indian major carps. It was revealed that 45% of scorable RAPD bands were specific to each species. Figure 4.4.2 presents RAPD profiles generated with one RAPD primer for 3 individuals from every species of Indian carps. You can see in the figure that these RAPD profiles are species-specific.

Figure 4.4.2. RAPD profiles generated with primer OPY10 for four species of Indian major carps; lines 1-3 – rohu (*Labeo rohita*), lines 4-6 – kalbasu (*L. calbasu*), lines 7-9 – catla (*Catla catla*), and lines 10-12 – mrigal (*Cirrhinus mrigala*) (from Barman et al. 2003, Aquaculture 217:115-123; Copyright 2003, reproduced with permission from Elsevier).

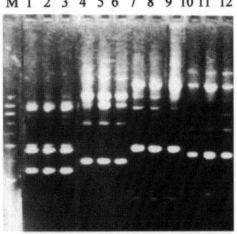

Microsatellite DNA markers are also useful for identification of species and distant hybrids. As mentioned above, microsatellites are usually highly variable with each locus containing large numbers of alleles. However, for species identification microsatellite loci with low variability are used. The locus should be monomorphic in every species (only one allele is present) and allelic difference should be observed between species.

Mia et al. (2005) have used variability for three microsatellite loci for detection of hybridization between two species of Chinese carps: silver carp (*Hypophthalmichthys molitrix*) and bighead carp (*Hypophthalmichthys nobilis*). For all three microsatellite loci (*Hmo1, Hmo 3 and Hmo 11*) species-specific alleles have been determined. Figure 4.4.3 demonstrates allele variation for three microsatellite loci. You can see that bighead and silver carps have different alleles while hybrids between two species have both species-specific alleles (as was noted above microsatellites are codomonant markers). The authors (Mia et al. 2005) noted that some analyzed fish displayed a mixture of both heterozygous and homozygous genotypes for different loci. This indicated that these fish are not F_1 hybrids but possibly backcross or F_2 hybrids. This study showed high effectiveness of application of microsatellites for detection of hybridization. Many individual fish, which have been identified morphologically as pure species, had microsatellite alleles of another species.

Figure 4.4.3. Example of allele variation for three microsatellite loci (*Hmo1, Hmo3* and *Hmo11*) in pure bighead carp, pure silver carp and hybrids between two species (from Mia et al. 2005, Aquaculture 247:267-273; Copyright 2005, reproduced with permission from Elsevier).

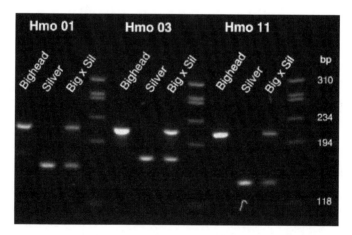

4.4.4. Parentage Determination

Microsatellites are the most popular type of markers for this purpose due to a large number of alleles per locus and codominant inheritance. High variability gives possibility to identify alleles, which are unique to specific fish parents, and using these data to identify parentage of each individual in mixed progeny. Studies on parentage determination using microsatellites have been performed in many fish species. (It should be noted that microsatellites are widely used for paternity testing in humans).

We will consider several examples on application of microsatellites for parentage determination in fish. Garcia de Leon et al. (1998) have analyzed parentage of offspring in mixed progeny of sea bass *Dicentrarchus labrax*. Three females and three males were genotyped for two microsatellite loci (Labrax-3 and Labrax-13). Genotypes of these parental fish are shown in the Table 4.4.3. Progenies have been artificially produced according to a scheme 3x3; nine possible progenies have been obtained. A mixture of all progenies has been reared in a single tank. The genotypes of 786 fish from the mixed progeny for these two microsatellite loci were determined. On the basis of these data the parentage (i.e. identification of male and female parents) of each individually analyzed fish was determined.

Table 4.4.3. Genotypes for two microsatellite loci of sea bass parents used in crosses (from Garcia de Leon et al. 1998; Aquaculture 159:303-316; Copyright 1998, reproduced with permission from Elsevier).

Locus	M1	M2	M3	F1	F2	F3	Number of alleles
Labrax-3	182/162	176/154	144/124	162/124	144/128	178/138	9
Labrax-13	176/160	168/164	154/150	192/142	168/158	170/166	11

F-female; M-male.

You can see in Table 4.4.3 that variability of sea bass parents for these two microsatellite loci was very high. The possible maximum number of alleles for 6 parent fish is 12; in this ideal case every fish parent would have two unique alleles. In real situation presented in Table 4.4.3 for locus Labrax-3 the number of alleles is

9; some alleles are unique (182, 176, 154, 128, 178 and 138) while 3 alleles (162, 144 and 124) are repeated in two fish parents. For locus Labrax-13 the number of alleles observed in fish parents is 11; in this case all recorded alleles besides allele 168, which was observed in one male (M2) and one female (F2) are unique. Therefore, locus Labrax-13 is more powerful than locus Labrax-3 for parentage determination. As was mentioned above, based on combined data on variability for two microsatellite loci in offspring the parentage of every fish from mixed progeny was successfully determined.

Hara and Sekino (2003) analyzed parentage of offspring in the case of natural communal spawning in Japanese flounder (*Paralichthys olivaceus*). Six females and eight males were placed in a tank for communal spawning and resulting mixed progeny was raised. Parentage of every fish in a sample (n=139) from mixed progeny was determined using variability for four microsatellite loci. These four alleles were highly variable: the number of alleles per locus observed in fish breeders varied from 17 to 21 and every allele had 11-12 unique alleles (which were found only in one individual breeder). Analysis of parentage showed that from the 14 fish breeders, which were stocked for spawning, only 8 (4 females and 4 males) contributed to the production of offspring.

The possibility of determining parentage of individual fish in mixed progenies is a very important tool for fish selection. As will be shown in Chapter 5, one of the main methods of fish selection is family selection. Family selection presumes simultaneous raising and comparison of a large number of families. Families should be raised separately until fish reach size when they can be individually marked (tagged). Separate rearing requires several replicates for each family (in order to diminish the influence of environmental factors). Parentage identification of fish using microsatellites can revolutionize the process of selection. In this case it is possible to raise all families together (under the same conditions) from the beginning (as was shown above in examples with sea bass and Japanese flounder).

Microsatellites are used also for determination parentage in fish inhabiting natural waters. It is powerful tool to analyze complex spawning systems known in fish. For example, Neff (2001) used microsatellites for paternity analysis and breeding success in bluegill sunfish (*Lepomis macrochirus*). Bluegill males demonstrate different types of reproductive behavior. The larger "parental" males construct nests, court females and provide parental care for offspring. By contrast, so-called "cuckolder" males do not build nests but steal fertilizations in the nests of parental males. There are two types of cuckolder males, which demonstrate different behavioral tactics. Smaller "sneakers" hide near the nest and wait for the moment when female releases eggs. Larger "satellite" males mimic coloration and behavior of females and try to deceive parental males that several females are present in the nest. The analysis of microsatellite markers showed (Neff 2001) that in the investigated colony the actual paternity of parental males was about 80% while 20% of offspring originated from cuckolder males.

Single nucleotide polymorphism markers (SNPs) can also (along with microsatellites) be used for parentage determination in fish. The disadvantage of SNPs for this application is that each locus has only two alleles but this may be compensated for by a large number of analyzed loci. The designed computerized model (Anderson and Garza 2006) shows that genotyping for 60-100 polymorphic SNP loci allows identification of parentage in cases when thousands fish parents are involved.

4.4.4. Genetic Mapping

Molecular markers are powerful tools for linkage analysis and genetic mapping in fish (as well as for other cultivated animals). Microsatellites, RAPDs, AFLP markers and SNPs are usually used for this aim. The first genetic map of DNA markers in fish was compiled for the model species zebrafish (Postlethwait et al. 1994). Later, genetic maps of molecular markers were compiled for many aquaculture species including tilapia, rainbow trout, and channel catfish.

Usually before linkage analysis, the level of heterozygosity in a fish genome is increased by means of intraspecies (crossing of lines or strains within species) or interspecies (crossing of different species) hybridization. The classical method of

98

genetic analysis of linkage and genetic mapping is usually used for creating genetic maps. Parents used in crosses are genotyped for many molecular marker loci. According to genetic segregation in progenies groups of linkage are identified, and the genetic distance between loci is determined.

For example, Agresti et al. (2000) compiled a genetic linkage map using microsatellites and AFLP markers for hybrid tilapia (*Oreochromis niloticus* x *Oreochromis aureus*) (Figure 4.4.4). The 214 segregating markers (60 microsatellites, 154 AFLPs) loci were placed into 24 linkage groups which correspond to 24 chromosomes in a haploid set. However, the authors (Agresti et al.

Figure 4.4.4. Linkage map of the hybrid male tilapia; microsatellites are identified by 'UNH' and shown in bold; AFLPs are identified by G*CNNN (from Agresti et al. 2000, Aquaculture 185:43-56; Copyright 2000, reproduced with permission from Elsevier).**

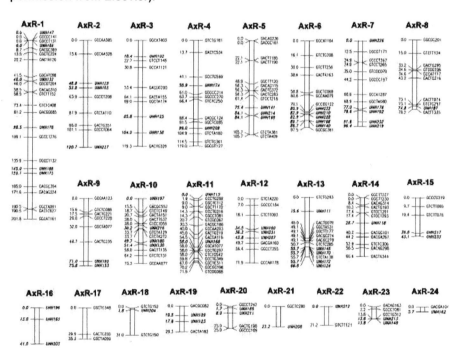

2000) noted that these linkage groups cannot be assigned to certain chromosomes in a karyotype. As will be shown in Chapter 7, the important task of fish genomics is an integration of genetic linkage maps with high resolution chromosomal physical maps obtained by DNA sequencing.

4.4.5. Identification of Quantitative Trait Loci

Using genetic maps of molecular markers, it is possible to determine locations of genes controlling important quantitative traits. These locations are called "quantitative trait loci" or QTL. QTL are identified by analysis of association between variability for some quantitative trait with segregation of DNA markers. Currently the identification of QTL is an intensively developing direction in genetics (genomics) of aquaculture species. Data on QTL for many quantitative traits have been obtained in rainbow trout, Atlantic salmon, tilapia and some other species.

For example, Sakamoto et al. (1999) identified quantitative trait loci (QTL) for spawning time in rainbow trout. The association between spawning time of females in backcross progeny between spring and fall spawning strains and segregation of microsatellite markers has been analyzed. The performed analysis has revealed association between some microsatellite markers with this quantitative trait. Identified QTL are located in five linkage groups (LG) 3, 8, and 23 (Figure 4.4.5).

Figure 4.4.5. Linkage groups possessing spawning time QTL (approximate locations indicated with a hatched bar) (Sakamoto et al. 1999, Aquaculture 173:33-43; Copyright 1999, reproduced with permission from Elsevier).

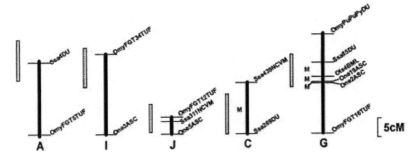

4.4.6. Marker-Assisted Selection

Marker-assisted selection (or MAS) is another possible practical application of molecular markers. The main idea of this method is that identified molecular markers associated with desirable phenotypes may be used as "express" criterion for selection of individuals for further breeding. Marker-assisted selection should be based on development of high-resolution linkage maps, identification of QTL and analysis of association between specific markers and expression of given quantitative trait. The significance of this method has been shown for plants and agricultural animals; the first data in this area have been obtained for aquaculture species. The potential benefits of marker-assisted selection may be especially useful for traits that are difficult or expensive to measure.

4.4.7. Identification of Genetic Sex

Molecular markers may be used for identification of genetic sex, i.e. for identification of composition of sex chromosomes in karyotype of given individual. This is possible when specific molecular markers are located on sex chromosomes and/or linked with sex-determining genes.

One of the first systems to identify genetic sex in fish was elaborated by Devlin et al. (1994a) for chinook salmon (*Oncorhynchus tshawytscha*). This species has male heterogamety: males have XY formula of sex chromosomes, females - XX. The probe for specific DNA region in Y-chromosome have been isolated and sequenced. Based on this sequence, primers were designed and used in polymerase chain reaction (PCR) to amplify genomic DNA. PCR analysis yielded a 209 bp fragment that was specific to males. Profiles of PCR amplified products are given in Figure 4.4.6. The squares in this figure represent males while the circles represent females. You can see in the figure that the parental male and all 5 male offspring have the Y-chromosome-specific fragment (209 bp). Lanes 1-3 represent the profiles of phenotypic males from androgen-treated group. This group consisted of normal genotypic males XY, and sex-reversed males XX. According to DNA profiles it is possible to identify the genotype of males in this progeny without performing test-

Figure 4.4.6. Profiles of PCR amplified products from chinook salmon parents with 5 sons and 5 daughters and 3 fish androgen-treated group (lanes 1-3) (from Devlin et al. 1994; Aquaculture 128:211-220; Copyright 1994, reproduced with permission from Elsevier).

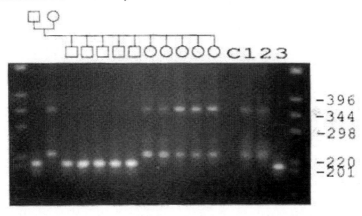

crosses with normal females. It may be seen that males 1 and 2 have genotype XX, while male 3 has genotype XY. (Hormonal sex reversal in fish is described in Chapter 6.)

Besides the main directions described above, DNA markers are used for other more specific tasks. For example, DNA markers can be used in studies on induced gynogenesis in fish to prove exclusion of paternal inheritance (Chapter 6).

Chapter 5. Selective Breeding and Hybridization

5.1. Introduction: Methods of Genetic Improvement of Fish

Chapters 5-7 of the book contain materials on methods of genetic improvement of fish, i.e. production of fish with desirable characters. All possible methods of genetic improvement of fish may be divided into three basic groups.

The first group includes the following classical methods of genetic improvement of animals:

- **Selection**
- **Crossbreeding or intraspecies hybridization**, i.e. crossing of fish from different strains, lines, or populations from the same species
- **Interspecies hybridization**, i.e. hybridization of fish from different species

These methods are described in Chapter 5. Also Chapter 5 includes **inbreeding** as a specific form of breeding.

The second group is so called methods of chromosome set and sex manipulation in fish. These methods are:

- **Induced gynogenesis**
- **Androgenesis**
- **Induced polyploidy**
- **Hormonal sex reversal and genetic sex regulation**

All these methods are described in Chapter 6.

Finally, the third group of methods for genetic improvement of fish includes **gene engineering** and production of **transgenic** fish. Production and properties of transgenic fish are presented in Chapter 7. Also, Chapter 7 includes materials on **genomics**.

5.2. Fish Selection

5.2.1. Artificial Selection and Its Forms

Selection is the process that determines which individuals become parents and how many offspring they produce. Under natural conditions there is **natural**

selection, which means that better fit individuals have better opportunities to survive and to leave more numerous progeny. Selection performed and controlled by man is called **artificial selection**. Three forms of artificial selection are distinguished:

- **Directional**
- **Stabilizing**
- **Disruptive**

Directional selection is most common form of selection. In this case, among animals of the parental generation, individuals having the best values of trait (according to the selectional goal) are selected for breeding. As a result of directional selection, the mean value of the trait shifts towards the chosen direction. Directional selection is a basic form of artificial selection. It is the main method for genetic improvement of animals.

In the case of **stabilizing** selection the animals with values of trait close to the mean value are selected for breeding. This results in decreasing variability for this trait in the offspring but the mean value of trait remains the same. Stabilizing selection in aquaculture may be used for increasing fish fitness for a certain standard technology. For example, for artificial spawning of fish it is desirable to decrease variability of females according to their reaction to the injection of gonadotropic hormones for induction of final oocyte maturation and ovulation.

In the case of **disruptive** (or bi-directional) selection, individuals with extreme values of some trait are bred separately, while individuals with the mean value are culled. Disruptive selection is usually used for experimental purposes to investigate response to selection performed in two different directions (plus and minus selection). Disruptive selection may also be used for practical application, for example, for development of lines with early and late spawning time during the spawning season. This makes it possible to extend the spawning season.

5.2.2. Peculiarities of Fish as Objects of Selection

The main peculiarities of fish as objects of selection are listed below:

1. Almost no aquaculture species are domesticated, a few are at initial stages of domestication. There is very big difference in this respect between fish (as well as other aquaculture objects) and main agricultural animals, which passed through a long-term process of domestication and selection. It is considered (Gjedrem and Baranski 2009) that less than 10% of world aquaculture production is based on genetically improved stocks. It is a fairly common practice in aquaculture to collect fry for rearing (or fish breeders for spawning) in natural waters.

However, it needs to take into account that even if fish are cultivated in captivity during a relatively short period, they are under permanent "domestication" pressure. Selection for fitness under new conditions (ability to survive under high density, ability to consume artificial diet etc.) occurs. It is very demonstrative, for example, when juveniles of largemouth bass are trained to consume an artificial diet. Up to 30-40% of fish do not take artificial diet and therefore die.

2. Fish usually have much higher fecundity than agricultural animals. High fecundity of fish results in both advantages and disadvantages. High fecundity permits one to conduct very intensive selection. On the other hand, one needs to be very cautious about inbreeding (crossing of relatives), which can result in loss of variability and inbreeding depression.

3. External fertilization. Together with high fecundity external fertilization gives the possibility to perform many crosses at the same time. Also, the opportunity to influence gametes and early embryos makes it possible to develop such specific methods as induced gynogenesis and polyploidy.

4. Not individual but group indices are recorded. In aquaculture practice total production from the square or volume units are recorded (for example kilograms per hectare). This is a big difference from agricultural animals where individual indices are used (for example, the volume of milk from one cow, carcass weight for one animal, number of eggs per one hen). In this respect, aquaculture has similar features with plant breeding. (Fish juveniles for stocking are sometimes even called seed stock.) Because of small size and large quantities it is relatively difficult to tag and record fish individually (especially early age groups), while the probability of stock contamination is very high.

5. Inhabitance in water. Because of this factor it is more difficult to observe animals and to record data (for example, on fish feeding in ponds). Also it is more difficult to standardize conditions of fish rearing. Ponds are very different according to their hydrobiological and hydrological regimes. Therefore for reliable comparison different fish groups should be raised in many replicates.

6. High dependence of fish on temperature and other environmental factors. Fish are poikilothermic animals (i.e. they have the temperature of the surrounding water), whose growth and development tightly depend on temperature. Other factors such as quantity and quality of feed, stock density have a great influence on fish growth. High variability caused by environmental conditions makes it more difficult to reveal genetic differences between groups of fish.

5.2.3. Directions and Objectives of Fish Selection

In this lecture the directions and objectives of fish selection are presented. The traits, which already have been used in practical selection, or those having only potential significance for genetic improvement of fish, are briefly described.

1. Selection for growth rate improvement

The feed conversion efficiency, which is determined as a ratio of amount of consumed feed to weight gain, is a very important trait for domestic animals. However, it is difficult to record this trait in fish due to inhabitance in water and communal rearing (especially when fish are reared in ponds). Therefore, growth rate is used as the most important trait in selection programs with aquaculture species. In contrast to feed conversion efficiency, growth rate may be easily determined by measurement of fish body weight (or length). Absolute growth rate is determined as the increase of body weight during a certain period of time.

Theoretically, better growth rate of fish may be achieved by two ways: a) by more effective feed conversion, i.e. by decreasing the amount of consumed feed to weight gain, or b) by increasing the amount of feed consumed by fish, i.e. by increasing appetite. In the last case, fish will be able to reach market size more quickly but the efficiency of feed conversion remains at the same level.

In farm animals, a high correlation between growth rate and the level of feed conversion was revealed. In some fish species this correlation was observed also. According to Thodesen et al. (2001) for Atlantic salmon the correlation coefficient between feed efficiency ratio and growth rate was close 0.80. High correlation between these two traits means that selection for growth rate results indirectly in improving feed efficiency since fast growing fish have better feed conversion.

The results of selection show that improved growth in fish can be achieved by different ways. Knibb et al. (1998) have reported that better growth rate of offspring from selected for growth rate gilthead sea bream (*Sparus aurata*) was caused by better feed conversion ratio (1.73 vs. 1.92 in offspring from unselected fish) while the amounts of feed consumed by fish of two groups were the same. Li et al. (2001) and Bosworth et al. (2007) have reported that the growth superiority of recently developed channel catfish strain NWAC103 in ponds was due primarily to increased feed consumption. Thodesen et al. (1999) have shown that higher growth rate of Atlantic salmon from selected line compared with wild line was the result of both factors: greater feed consumption and more effective feed utilization for growth. Offspring from salmon selected for increased growth rate for five generation consumed 40% more food and had 20% better feed conversion ratio than progeny from wild-caught fish.

Selection for growth rate can result in some undesirable changes. Knibb et al. (1998) have recorded that selection of gilthead sea bream (*Sparus aurata*) for increase of body weight resulted in significant increase of gonadosomatic index (GSI, ratio of gonad weight to whole body weight) compared to unselected (control) group. Quinton et al. (2005) have indicated that selection for larger harvest body weight in Atlantic salmon resulted in undesirable increase in flesh fat content.

As indicated in Chapter 3, fish weight and length are quantitative traits. Therefore, growth rate of fish depends on environmental conditions such as water temperature, quality and quantity of feed, etc. It was also noted that the heritability of fish weight is low (usually 0.1-0.4). In order to evaluate genetic differences in growth rate between different fish groups one needs to minimize influence of environmental factors. First of all, it may be done by rearing fish under the same conditions. Since ponds or tanks are different, in the case of separate rearing each group of fish should be raised in several replicates.

Environmental factors can also influence the within-population variability of fish. The Japanese scientists N. Nakamura and S. Kasahara in a series of studies (1955-1961) with common carp have showed that competition for feed or poor feed quality can drastically increase fry variability by appearance of so called "shoot fish" (or "jumpers"), i.e. individuals, which have superior growth rate as compared with most fish in a group. The results of one demonstrative experiment performed by Nakamura and Kasahara (1956) are given below. Two groups (A and B) of common carp larvae of the same origin were fed with live zooplankton collected by different methods. Group A was fed with small forms of plankton, which had passed through a 400 micron mesh screen while group B was fed with large zooplankton left on the screen with adding only minute portion of small zooplankton. The length distributions of fry in these two groups after 20-day rearing are shown in Figure 5.2.1. You can see in the figure that these two groups differed according to fish length distribution. Fish distribution in group A was almost symmetrical and low variability was observed while the distribution curve of group B was skewed drastically to the right because of appearance of "shoot" fish (see Figure 5.2.1). The authors (Nakamura and Kasahara 1956) explained that the small size and uniformity of the food particles given to group A provided an equal opportunity for all larvae to feed. Therefore fish in this group had relatively uniform body length. In group B, fed with larger feed particles, only a relatively small number of larger individuals, that were able at the beginning to consume large particles, received sufficient food to grow well. The difference between larger individuals and smaller ones rapidly increased and the "shooting" phenomenon occurred. The results of this (and similar experiments) clearly demonstrate that superiority of "shoot" individuals or "jumpers" does not have genetic basis but is caused by environmental factors.

Figure 5.2.1. Body length distribution of two batches of carp fry reared with food of different particle size (from Nakamura and Kasahara 1956; reproduced with permission from The Japanese Society of Fisheries Science).

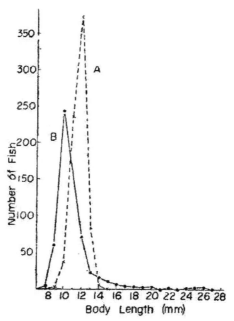

In predatory fish "jumpers" are usually cannibalistic individuals which start consuming smaller fish from the same cohort instead of artificial diet. For example, appearance of extremely large, cannibalistic individuals is a common phenomenon when largemouth bass fry are fed an artificial diet.

Another factor that makes the comparison in growth rate between groups of fish more difficult is the difference in initial weight. The difference in weight is multiplied along with fish growth. For example, Wohlfarth and Moav (1972) have shown that a difference of 1 g of initial weight at stocking of 30-50 g common carp gave a difference of 3-4 g of final weight at harvest of 500-700 g fish. The special indices are usually calculated (for example, specific growth rate or growth rate index) in order to compare growth rate of groups of fish, which had different initial weight.

2. Selection for resistance to diseases

Selection for resistance to diseases is based on between-population and within-population genetic variability for this trait observed in many fish. Several long-term programs on selection of fish for resistance to diseases have been performed up to the present. Examples:

1. Selection of brook trout (*Salvelinus fontinalis*) and brown trout (*Salmo trutta*) for resistance to furunculosis in USA (Ehlinger 1977);

2. Selection of common carp for resistance to dropsy in Russia (Kirpichnikov et al. 1993);

3. Selection of Atlantic salmon for resistance to infectious pancreatic necrosis in Norway (Gjedrem 2000; Storset et al. 2007).

3. Selection for improvement of reproductive traits

As a rule, fish have high fecundity. Therefore, selection for increase in fecundity is not so important. However, several other traits connected with reproduction have certain significance. One trait is the time of spawning during spawning season. Frequently, fish have a very short period of spawning. Sometimes it is important to spawn fish beyond their natural spawning time. This would make it possible to supply production systems with fingerlings over a longer period or extend the growing season by earlier spawning. Several long-term programs selecting for earlier spawning in several species have been successful.

Another trait, which may have potential significance for selection, is the age of sexual maturity. For fish with slow maturation, such as sturgeons, selection for speeding of this process will be useful. But usually the opposite direction, later age of sex maturation, is more desirable. It is known that the development of gonads in fish causes retardation in the somatic growth rate. Therefore, it is desirable if maturation would be reached after achievement of market size. It will be a significant advantage for such aquaculture species as common carp and tilapia.

4. Selection for changes in morphological, physiological and biochemical traits

Selection for these traits may change body shape, improve quality of flesh etc. A good example of selection for a morphological trait is the change of body shape in common carp during domestication and selection. In common carp one of the main traits, indicating the rate of domestication is the relative height of the body. Wild fish have elongated body; during domestication the relative height of the body has increased. Usually fish with higher body have better growth rate and, therefore, higher productivity. Besides this, fish with a higher body have a larger proportion of edible parts.

For common carp such traits as content of fat in flesh or the number of intramuscular bones may be important. Investigations recording variability of these traits have been already performed.

For rainbow trout and salmon one of the possible tasks of selection is to intensify flesh color. It may be reached by selecting fish with an increased content of carotenoids in the flesh.

5. Selection for fitness under certain conditions.

The aim of this direction is the development of fish which are capable of being raised under specific (usually more severe) conditions for a given species. For example, it is desirable to develop tilapia strains resistant to cold temperature. One cold resistant breed of common carp ("Ropsha") has already been developed in Russia (see part 5.4.1).

6. Selection according to selection indices

This direction is popular for agricultural animals, but in fish selection only several attempts in this direction have been made. In the case of selection indices, several different traits are taken into account at the same time (for example, growth rate, content of fat, viability). The sum of the points (selection indices) for each trait is used for a final complex evaluation of the fish.

5.2.4. Methods of Fish Selection

There are two main methods of artificial selection with regard to how it is used for the evaluation of animals:

- **Individual selection**
- **Selection based on relatives**

5.2.4.1. Individual or Mass Selection

In the case of **individual selection**, animals (individuals) are selected with regard to their phenotypic values only (phenotypic selection). It is assumed that individuals having better phenotypic performance also have better genetic breeding values. Individual selection is the simplest form of selection, but its effectiveness has been proven repeatedly in fish selection. The whole process of domestication is the result of individual selection.

In animal breeding, the term **mass selection** is often used for individual selection when selected individuals are put together for spawning (Falconer and Mackay 1996). As we know, in fish breeding mass spawning (natural or induced) is frequently used. Therefore in aquaculture terms 'individual' and 'mass' selection are used as synonyms.

The general scheme of mass (individual) artificial selection is given in Figure 5.2.2. You can see in the figure that individuals from the parental population with values of a trait that exceed a certain value (**truncation point**) are chosen for breeding. As a result of selection, the mean value of a trait shifts in a desirable direction.

The effectiveness of selection per one generation may be determined according to a formula (Falconer and Mackay 1996):

$$R = h^2 S$$

where **R** is effectiveness (response) of selection, i.e. the difference in the mean value of a trait in the offspring and the parental generations (see Figure 5.2.2); h^2 is heritability and **S** is selection differential, i.e. the difference in the mean value of a trait in selected individuals and the mean value in the whole population before selection (see Figure 5.2.2). (Reminder: this formula was presented in Chapter 3 for calculation of realized heritability).

Another index, **selection intensity** (*i*) is equal to the selection differential (S) expressed in terms of standard deviation:

$$i = S/\sigma$$

Figure 5.2.2. General scheme of mass (individual) artificial selection.

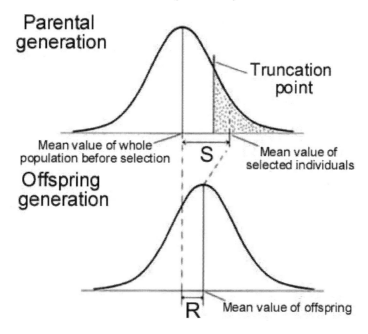

Then:

$$R = ih^2\sigma$$

There is dependence between intensity of selection (*i*) and proportion of selected individuals (*p*), which is determined by the rules of a normal distribution. (Sometimes the special term "selection severity" is used for designation of proportion of selection individuals). This dependence is shown in Table 5.2.1. According to this dependence, for example, if proportion of selected individuals is 10% (i.e. 1 from 10 fish is selected) intensity of selection is 1.76; if proportion of selected individuals is 1% (1 fish from 100 is selected) intensity of selection is 2.67.

Table 5.2.1. Dependence between proportion of selected individuals (p) and intensity of selection (*i*).

p, %	30	25	20	15	10	5	1	0.5	0.1	0.05
i	1.16	1.27	1.40	1.55	1.76	2.06	2.67	2.89	3.37	3.55

According to the formula $R=ih^2\sigma$, the effectiveness of selection (R) may be increased by increasing the intensity (*i*) of selection. Because of the high fecundity of fish, the intensity of fish selection is much higher than that applied in selection of agricultural animals. Intensity of fish selection may reach 3.37 when proportion of selection individuals equals 0.1% (i.e. 1 fish is selected from 1,000). However, it is preferred that the intensity of fish selection not be so high for several reasons:

- The appearance of 'shooting fish' or 'jumpers' resulting from environmental factors;
- The appearance of correlated negative changes resulting from too intense selection. For example, selection for body weight with excessive intensity may result to selection of fish with undeveloped gonads, decreased viability etc.

It is considered (Kirpichnikov 1981) that intensity of mass selection in fish should not be larger than 1.5-2.0; only in extreme cases it can reach 2.5. The optimal proportion of selected fish is 5-10%.

In the general scheme of selection presented in Figure 5.2.2 the response of selection (R) is determined as the difference between the mean values of a trait in offspring and parent generations. By this "classical" method the effectiveness of selection is determined in agricultural animals. However, this method is not usually used in selection of fish (and other aquatic animals). In aquaculture practice it is difficult to standardize conditions of fish rearing in two consecutive generations. There may be differences in temperature conditions, type of feed, stock density etc. Therefore, in aquaculture practice another method for determination of selection response is used. The scheme of this method is given in Figure 5.2.3.

Figure 5.2.3. Scheme for determination of response of selection in aquaculture practice.

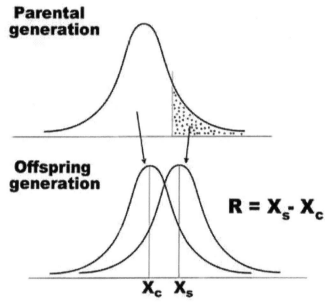

According to this method, besides a selected group of animals (with the best values of trait) a control group of animals is taken randomly from the initial population. Offspring, produced from the selected and control groups, are reared under similar conditions and with several replicates. The response of selection (R) is determined as the difference in mean values of the trait between the selected (X_s) and control (X_c) groups as $R = X_s - X_c$.

As an example, we will consider an experiment using mass selection for weight of the catarina scallop performed by Ibarra et al. (1999). In Figure 5.2.4 the weight distributions of animals in the whole parental population and in the selected group are shown. The 50 heaviest animals from 480 animals have been selected; proportion of selected individuals was 10.42% (50/480), selection differential was 18.28 g (see Figure 5.2.4). Along with this selected group, the control group was formed by randomly taking other 50 animals from the population. Progenies obtained from the selected and control groups were reared under similar conditions. Mean weight of offspring obtained from the control group of breeders was 47.42 g. Mean weight of offspring obtained from selected breeders was 55.21 g. According to the formula presented above, the response of selection (R) was determined as 55.21 - 47.42 = 7.79 g. Realized heritability (h^2) was calculated as R/S or 7.79/18.28 = 0.43.

Figure 5.2.4. Weight distributions in parental population and in selected group of catarina scallop* (*Argopecten ventricosus*) (from Ibarra et al. 1999, Aquaculture 175:227-241; Copyright 1999, reproduced with permission from Elsevier).

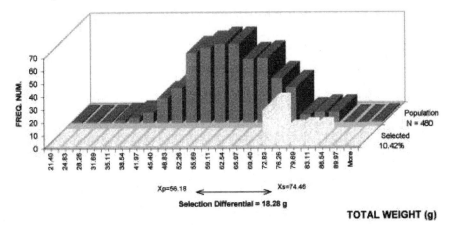

Another example is a study on mass selection for body weight in channel catfish performed by Dunham and Smitherman (1983). (This study was briefly considered in Chapter 3 when the method of realized heritability calculation was

---

* This book is on fish genetics. However, for some universal subjects (e.g. application of DNA markers or artificial selection) demonstrative examples using invertebrate aquaculture species are presented.

explained.) From three strains of channel catfish 10% of fish having the largest weights were chosen as the selected groups. Random samples of fish were taken for respective control groups. Progenies obtained from selected and control brood fish were reared under similar conditions with several replicates. Results of this selection experiment are given in Table 5.2.2. You can see in the table that responses of selection were determined as the difference between mean weight of progeny from selected fish (S) and mean weight of progenies from randomly taken fish (R). The pooled data for all 3 lines were as follows: response of selection = 64 g; selection differential = 190 g; calculated realized heritability was 64/190 = 0.34 (see Table 5.2.2).

Table 5.2.2. Response to selection for body weight in three channel catfish strains (Rio Grande, Marion and Kansas) grown in earthen ponds (S = select; R = random) (from Dunham and Smitherman 1983, Aquaculture 33:89-96; Copyright 1983, reproduced with permission from Elsevier).

Population		Mean weight (S.D.) (g)	N	Response (g)	Selection differential (g)	Realized heritability ± S.E.
Rio Grande	(S)	431 (136)*	1044	63	263	0.24 ± 0.06
	(R)	368 (109)	1068			
Marion	(S)	486 (146)*	1674	73	145	0.50 ± 0.13
	(R)	413 (99)	1656			
Kansas	(S)	513 (142)*	1692	54	163	0.33 ± 0.10
	(R)	459 (107)	1764			
Pooled	(S)	477 (156)*	4410	64	190	0.34 ± 0.07
	(R)	413 (117)	4488			

*Significantly different ($P < 0.001$).

In the examples presented above, the effectiveness of mass selection performed in one generation was described. Below, the effectiveness of mass selection in several long-term selection programs in fish is shown:

1. Selection of rainbow trout for early spawn date. Siitonen and Gall (1989) reported results of a long-term selection program for early spawning in rainbow trout. Six consecutive generations of mass selection were performed in the period 1972-1983. Selection was effective; the average response of selection was about 7 days per generation. The change in mean spawn date during the selection process can be seen in Figure 5.2.5.

Figure 5.2.5. Mean spawn date over six generations of selection for early spawn date in two year-classes of rainbow trout broodstock. The first year of each year-class was generation zero. Mean date refers to number of days from December 31 (from Siitonen and Gall 1989, Aquaculture 78:153-161; Copyright 1989, reproduced with permission from Elsevier).

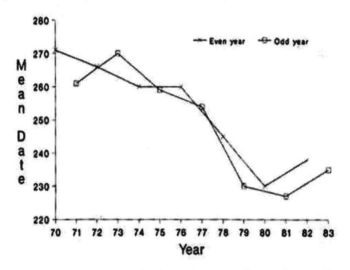

2. **Selection of silver barb for improvement of growth rate.** Hussain et al. (2002) have used mass selection for improvement of growth rate in silver barb (*Barbodes gonionotus*). Three generations of mass selection were performed. Percent weight gain differences between the selected and control groups are shown in the Figure 5.2.6. The weight gain values of the third generation of the selected group showed 22% superiority over the nonselected control group.

3. **Selection of common carp for resistance to dropsy.** Dropsy is a severe infectious disease of common carp in the southern region of Russia. A long-term selection program for resistance to this disease has been performed; the program was started in the middle of 1960s (Kirpichnikov et al. 1993). In 1979-1980 the comparative resistance of fish from four generations of mass selection (from 2nd to

Figure 5.2.6. Percent weight gain difference in selected groups of silver barb over nonselected control groups in 3 generations of selection (from Hussain et al. 2002, Aquaculture 204:469-480; Copyright 2002, reproduced with permission from Elsevier).

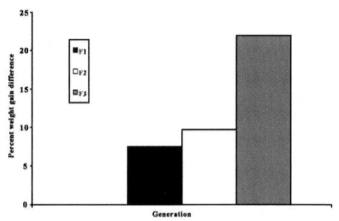

5th) was tested (Ilyasov et al. 1983). Fish of all four generations used for comparison were obtained at the same time. This was possible since fish breeders from previous generations of selection were kept at the experimental fish farm. The comparison has shown that mass selection for resistance to disease was effective: fish survival consistently increased from 43% in the second generation to about 72% in the fifth generation while percentage of healthy fish consistently increased from 15% (2nd generation) to 63% (5th generation).

Development of new biotechnological methods opens some new opportunities for determination of selection effectiveness. In order to evaluate effectiveness of selection in GIFT strain of tilapia (Genetically Improved Farmed Tilapia), Khaw et al. (2008) have compared growth rate of fish produced with cryopreserved sperm taken from males of the base population (originated in 1991) with fish produced with native sperm from males of ninth generation of selection (originated in 2003). Total effectiveness of selection for the body weight after 120-day grow-out period was 89 g or 64%, or about 7% per one generation of selection.

5.2.4.2. Selection Based On Relatives

An individual's own phenotype is not the only source of information about its breeding value. Additional information may be obtained from an evaluation of relatives. In this case, by investigation of close relatives one tries to evaluate the genetic breeding value of individuals.

There are several types of selection based on relatives used in agricultural practice. Their description and possibilities of application in aquaculture are given below.

1. Evaluation according to pedigree data

In this case, an individual is evaluated based on pedigree data, i.e. records on performance and achievements of ancestors (parents, grandparents etc.). This method of selection is very important for many agricultural animals, as well as for horses and dogs. Until recently there were no pedigree records in aquaculture practice at all. Development of DNA markers technology makes possible the pedigree tracing in fish. In the previous part the parentage determination using microsatellites was described. The application of this technique during several consecutive generations would give us pedigree of the given fish. These records may help to avoid inbreeding (see part 5.3). However, these pedigrees cannot be used directly for selection purposes since they do not contain any data on achievements of ancestors.

Another reason why this approach has restricted application in fish selection is that, as already mentioned, in aquaculture not individual but group indices are important.

2. Progeny testing

In this case, breeders, the males and females, are chosen according to quality of progenies obtained by a special system of crosses. Males to be tested are crossed with the same females, the performance of obtained progenies is investigated, and the best males breeders are determined. The best females are determined according to an opposite scheme where different females are crossed with the same males. The main idea of progeny testing is, at the final stage, the best selected males are crossed with the best selected females. The 'progeny testing' method is popular in selection of agricultural animals but in fish selection it is rarely used. This method is time-consuming because final selection of the parents cannot be carried out until offspring are evaluated.

3. Family selection

In contrast to the two methods of selection based on relatives described above, **family selection** or **family-based selection** is a widely used aquaculture practice.

The basic principle of family selection is that a population is considered not as the aggregation of individuals (as in the case of individual selection) but as aggregation of different families, i.e. groups of related individuals. This difference in approach to population structure, typical for individual and family selection, is seen in Figure 5.2.7. It is supposed that analysis of separate families instead of the whole population will give more accurate information on genetic values of individuals.

Figure 5.2.7. Different approaches to population structure in the cases of individual (whole population) and family selection.

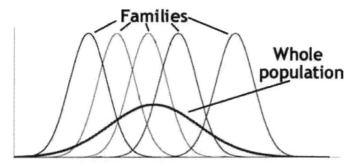

In animal selection, as well as in fish selection, one family is usually the progeny obtained by crossing one female with one male. Such families are called "full-sib families" since they consist of full siblings (sibs). Sometimes in fish selection families are obtained by crossing one female with two males (or *vice versa*). Such families are called "half-sib families".

Family selection has two basic forms: **between-family selection** and **within-family selection**. The third form, **combined family selection**, is the simultaneous application of the two above mentioned forms.

During between-family selection whole families are selected or rejected according to their mean phenotypic value. The general scheme of between-family selection is presented in Figure 5.2.8.

Figure 5.2.8. Scheme of between-family selection.

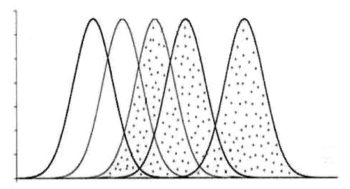

You can see in Figure 5.2.8 that from the 5 tested families, 3 have been chosen for further breeding. The next generation is produced by crossing females and males taken from different families in order to avoid inbreeding. In fish selection families are very numerous because of high fecundity. Therefore, only part of the fish is taken for further breeding. In the case of between-family selection, fish from each selected family are taken by random sampling (i.e. there is no within-family selection).

The main task of between-family selection is the proper evaluation of genetic values of different families. In order to decrease effect of environmental factors the compared family should be reared separately in tanks or ponds under similar conditions and with several replicates. The communal rearing of families can decrease the influence of environmental factors. However, in this case one needs to mark (tag) fish, which is possible only when fish reach relatively large sizes.

Effectiveness of between-family selection (R_f) is determined by the following formula (Falconer and Mackay 1996):

$$R_f = i\sigma_f h_f^2$$

where i is the intensity of selection, σ_f is the standard deviation of family means, and h_f^2 is the heritability of family means. It should be noted that the selection intensity of between-family selection will be less than that of individual (mass) selection since only a limited number of families are tested. Standard deviation also will be decreased since deviation of the means is less than deviations of individual values. However, heritability might be higher if all families are reared under the same conditions. Therefore, it is supposed that family selection might be more effective than individual selection when heritability of a trait is relatively small.

Longalong et al. (1999) used bi-directional (disruptive) between-family selection for frequency of early maturation in Nile tilapia. Forty two full-sib families were investigated according to the frequency of early maturing females. Distribution of families according to this trait is given in Figure 5.2.9. The 8 full-sib families with the highest frequency of maturation (HFM) and the 8 full-sib families with the lowest frequency of maturation (LFM) were used as parental broodstock for bi-directional selection for this trait (see Figure 5.2.9). The response to selection, measured as the difference between two progeny groups (+ and -), was highly significant.

Another form of family selection is **within-family selection**. The basic scheme of within-family selection is given in Figure 5.2.10. You can see that in this case all tested families are used for further breeding, but only best animals from each family are selected. Within-family selection may be successfully used either independently or in combination with between-family selection.

Effectiveness of within-family selection (R_w) is determined according to the following formula (Falconer and Mackay 1996):

$$R_w = i\sigma_w h_w^2$$

where *i* is the intensity of selection, σ_w is the standard within-family deviation, and h_w^2 is the within-family heritability of trait.

Figure 5.2.9. Distribution of tilapia families according to the frequency of early maturing females (from Longalong et al. 1999, Aquaculture 178:13-25; Copyright 1999, reproduced with permission from Elsevier).

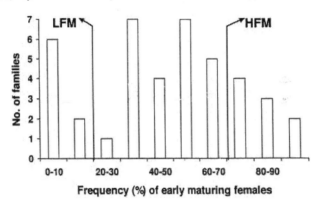

Figure 5.2.10. Scheme of within-family selection.

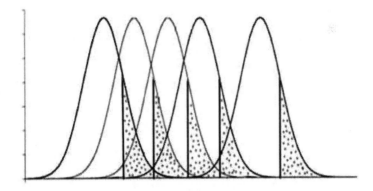

Fish from one family are under the same environmental conditions since from the beginning they are raised communally. Therefore within-family heritability of trait is usually high. Intensity of within-family selection may also be high due to possible large family size in fish. These factors increase the possible effectiveness of within-family selection. Application of only within-family selection (without between-family selection) gives one the possibility to perform family-based selection with limited resources (low number of ponds, tanks etc.).

The last form of family-based selection is **combined family selection**, i.e. the combination of between-family selection and within-family selection. The scheme of this method is given in Figure 5.2.11. In this case the best animals from the best selected families are used for further breeding. Effectiveness of combined family selection is determined as the sum of the effectiveness of the corresponding between-family and within-family components.

Figure 5.2.11. Scheme of combined family selection.

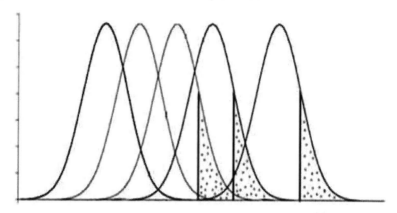

As an example of combined family selection we will consider a selection experiment by Hörstgen-Schawrk (1993) to improve growth rate in rainbow trout. Growth rate of full-sib families of rainbow trout was tested under two environmental conditions: low stocking densities in ponds and high density in tanks. Two generations of combined family selection were performed. The number of evaluated families varied between 13 and 19 for each of two selected lines (fall and spring) in each generation. On average, the best 5 families in both environments were selected among all families tested. Within each selected family the 15 to 16 individuals with the highest body weight were selected. All crosses within selection experiment were performed according to scheme which prevented mating of full sibs. For the evaluation of selection effectiveness, control groups compiled by random samples of fry drawn from all families, were raised in batches equal in size to families of the selection lines. Responses of selection varied for different lines and conditions. After 2 generations of selection the fish weight in selected lines was 7-27% higher than that in corresponding controls. Average realized heritability was 0.15.

The method of family selection was used in several long-term selection programs in fish. For example, Hershberger et al. (1990) reported the results of a 10-year selection and breeding program with coho salmon raised in marine net-pens. Four generations of selection were conducted; the results showed an average of 10% gain in fish weight per generation.

Long-term, large-scale selection program with Atlantic salmon based on the method of family selection was started in Norway at the beginning of the 1970th. As part of the National Breeding Program several hundred full-sib and half-sib families salmon were tested in each of four year classes. It is estimated that 10-15% improvement in fish growth rate was obtained per one generation of selection (Gjedrem 2000). By 2006, genetically improved Atlantic salmon of the seventh selection generation were disseminated to Norwegian farmers (Thodesen and Gjedrem 2006).

Recently, the method of family selection became more popular in fish selection. One of the most important reasons is the development of DNA genetic markers, namely microsatellites. As was mentioned above, during between-family or combined family selection, families should be reared in similar conditions with several replicates. This requires a large number of ponds or tanks. Communal rearing of many families together is very desirable since it diminishes the influence of environmental factors. Communal rearing, however, is limited by the methods of fish marking (tagging). Using standard methods only, fish of relatively large sizes may be successfully tagged. As shown in Chapter 4, application of DNA markers, usually microsatellites, gives possibility to identify parentage of individual fish in mixed progenies. Therefore, microsatellite technique may revolutionize the process of family-based selection by reducing drastically the requirement for ponds or tanks. Microsatellite technique has already been used in selection experiments in several fish species such as African catfish (Volckaert and Hellemans 1999) and channel catfish (Waldbieser and Walters 1999).

5.3. Inbreeding and Methods of Its Reduction

5.3.1. Inbreeding: Definition and Methods of Its Determination

Inbreeding is the mating of relatives, or individuals that are related to each other by ancestry. If two individuals have a common ancestor, both can carry **identical by descent** alleles, i.e. copies of alleles, which the ancestor had. When these relatives mate, part of their offspring will have identical by descent alleles in the homozygous state. Therefore, the main genetic effect of inbreeding is an increase of homozygosity (and correspondingly a decrease of heterozygosity).

Genomes of animals from natural or artificially cultivated populations are saturated with deleterious (i.e., lethal, semilethal, inducing decreased performance) recessive mutant alleles. The aggregate of these deleterious recessive alleles is called sometimes "genetic load". Since these alleles are recessive, they are not expressed in heterozygotes. Inbreeding transfers these deleterious alleles to homozygous state, and, therefore, they are expressed. The action of such alleles may result in **inbreeding depression**, i.e. a decrease in survival and productive traits of inbred animals.

Thus, the negative effect of inbreeding results from two phenomena: increase of homozygosity and occurrence of recessive deleterious mutations (genetic load).

The rate of inbreeding is measured by the **coefficient of inbreeding (F)**, which determines the probability of the increase of homozygosity in offspring. For example, if in an initial population 40% of the genes are in a heterozygous state, and F=0.20, the number of homozygous genes in the offspring increased up to 48%.

For domestic animals with pedigree records the coefficient of inbreeding can be calculated using special formulae based on the number of generations to a common ancestor.

The highest rate of inbreeding in one generation in plants and hermaphroditic animals is observed in case of self-fertilization (F=0.50). Among animals reproducing by the normal sexual mode, the maximum rate of inbreeding may be achieved by crossing of sibs or a parent with offspring (father x daughter or mother x son) (F=0.25). During successive generations of close inbreeding the coefficient of inbreeding increases but the rate of increase diminishes. Table 5.3.1 shows the value of coefficient of inbreeding for different types of crosses during successive generations.

Table 5.3.1. Value of coefficient of inbreeding for different breeding systems.

Breeding system	Generation						
	1	2	3	4	5	6......	10
Self-fertilization	0.50	0.75	0.875	0.938	0.969	0.984	0.999
Full-sibs mating or Parent x offspring	0.25	0.375	0.50	0.594	0.672	0.734	0.888
Half-sibs mating	0.125	0.219	0.305	0.381	0.449	0.509	0.691
Mitotic gynogenesis in fish	1.00	-	-	-	-	-	-

As you can see in Table 5.3.1, the complete homozygosity (F is about 1.0) is reached after 10 generations of self-fertilization. For comparison, I have also included mitotic gynogenesis in fish in Table 5.3.1. By this unique method, which is available in fish genetics, it is possible to reach complete homozygosity (F=1.0) in one generation. Description of mitotic gynogenesis is given in Chapter 6.

We have considered the method of determination of inbreeding coefficient in cases when relatedness of animals used in crosses is known from pedigree data. However, as a rule, pedigree records are not available in aquacultural practice. In these cases the coefficient of inbreeding is determined based on the number of animals used for breeding. This approach is based on the suggestion that in a closed population mating between relatives occurs by chance. It is obvious that the chance of mating between relatives increases with a decrease of population size.

Based on this approach, the coefficient of inbreeding for one generation is calculated according to the following formula (#1):

$$F = \frac{1}{2N_e}$$

where N_e is **effective breeding number** (or **effective population size**). Effective breeding number (N_e) is determined according to the following formula (#2):

$$N_e = \frac{4\,N_m{\cdot}N_f}{N_m + N_f}$$

where N_m and N_f are the number of males and females, respectively, used for breeding to produce offspring.

Combining the two formulae presented above, the following formula (#3) for determination of inbreeding coefficient can be deduced:

$$F = \frac{1}{8N_m} + \frac{1}{8N_f}$$

Dependence of coefficient of inbreeding on effective breeding number is shown in the Table 5.3.2. In Figure 5.3.1 the same dependence is given in graph form. You can see that value of effective breeding number (N_e) below 50 results in a drastic increase in value of coefficient of inbreeding (F).

Table 5.3.2. Dependence of coefficient of inbreeding on the effective breeding number.

Effective breeding number, N_e	Coefficient of inbreeding, F
2	0.25
4	0.125
6	0.083
8	0.065
10	0.05
20	0.025
50	0.010
100	0.005
200	0.0025
400	0.00125
500	0.00100

Figure 5.3.1. Dependence of coefficient of inbreeding on the effective breeding number.

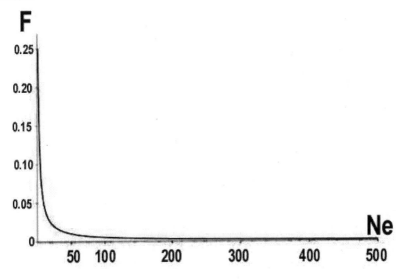

If the number of males and females used for producing of offspring is equal (i.e. $N_f = N_m$) then N_e is determined as the simple sum of the numbers of males and females: $N_e = N_f + N_m$. If the numbers of males and females are different N_e is determined according to formula #2 (presented above). In this case the rate of inbreeding depends mainly on the number of the less numerous sex. The basis of this dependence is obvious from formula #3 (presented above). As an example, the Table 5.3.3 presents the value of inbreeding coefficient based on using for breeding 10 females and varying numbers of males.

We have considered the calculation of coefficient of inbreeding for one generation. If coefficient of inbreeding is the same in each generation, the coefficient of inbreeding after t generations may be calculated according to the following formula (#4):

$$F_t = 1 - (1-F_x)^t$$

where F_x is coefficient of inbreeding for one generation determined based on formulae #1 or #3. Example: During 3 successive generations coefficient of inbreeding was 0.15 for each generation. Calculate the cumulative rate of inbreeding. Solution: $F_3 = 1 - (1-0.15)^3 = 0.39$.

Table 5.3.3. Value of coefficient of inbreeding (F) based on using for breeding 10 females and varying numbers of males (N_m).

Number of males, N_m	Coefficient of inbreeding, F
1	0.1375
3	0.075
5	0.0375
10	0.025
15	0.021
20	0.019
40	0.016
50	0.015
100	0.014

If coefficient of inbreeding is not constant from generation to generation its cumulative value can be determined using the average value of effective breeding number. Average N_e is calculated as **harmonic mean** (for definition see textbook on mathematics) according to the following formula (#5):

$$\bar{N}_e = \frac{t}{(1/N_{e1} + 1/N_{e2} + 1/N_{e3} + \ldots + 1/N_{et})}$$

where t is number of generations while N_{e1}, N_{e2}....N_{et} are effective breeding numbers in corresponding generations. Coefficient of inbreeding (F) for t generations is calculated as $1/(2 \times$ Average $N_e)$. Example: In 4 generation the values of effective breeding numbers were the following: N_{e1}=50, N_{e2}=50, N_{e3}=8, N_{e4}=50. Calculate average effective breeding number and coefficient of inbreeding for 4 generations. Solution: Average N_e = 4/(1/50 + 1/50 + 1/8 + 1/50) = 21.6; F_4 = 1/(2 x21.6) = 0.023.

According to formula #5 generations with the smallest effective breeding numbers have the greatest impact on the mean value of N_e. When N_e is very small in a certain generation, the population is said to have gone through a **bottleneck**. An expansion in numbers in further generations does not affect the previous inbreeding. It merely reduces the rate of new inbreeding.

5.3.2. Expression of Inbreeding Depression in Fish

In fish, as well as in other animals, inbreeding depression results in a significant decrease in productivity. In fish, inbreeding depression is usually expressed as:

- Decreased fish survival
- Increased percentage of deformed fish
- Frequently observed abnormalities in gonad development
- Decreased fecundity
- Decreased growth rate

Appearance of some of these signs indicates the possibility of inbreeding in fish stock.

Data on the direct effect of inbreeding on productive traits were obtained for several aquaculture species. Several examples are listed below:

- In rainbow trout after two generation of brother x sister mating the percentage of deformed fry increased by almost 2 times. At the same time fry survival decreased by 30%, and mean weight attained by fish at 364 days of rearing decreased for 33.5% (Kincaid 1976).
- In channel catfish two generations of brother x sister mating resulted in a 19% decrease in body weight (Bondari and Dunham 1987).
- Evans et al. (2004) investigated the effects of different levels of inbreeding on productive traits in Pacific oysters. For this purpose the oyster families with different coefficients of inbreeding (0, 0.0625 and 0.203) were created using special breeding schemes. The obtained results showed a decrease in different productive traits (survival, body weight and yield) along with an increase in the rate of inbreeding.

Numerous data on inbreeding depression in fish were also obtained when performance of gynogenetic progenies was investigated. It is known that induced gynogenesis results in an increase in homozygosity (Chapter 6).

5.3.3. Methods of Inbreeding Reduction

Basic methods of inbreeding reduction are listed below.

1. Use a sufficient number of brood fish. One needs to follow the rule: the larger effective breeding number (N_e), the better. Twenty-five to fifty pairs of breeders may be regarded as an acceptable minimum. As indicated in 5.3.1, a value of N_e below 50 results in a drastic increase in the rate of inbreeding.

One needs also to take into account that the real number of parents contributing to offspring can be less than the number of breeders placed for spawning. (As was shown in Chapter 4 in example with parentage determination in the case of mass spawning.)

2. Use equal numbers of females and males for breeding. As was shown in part 5.3.1, if numbers of males and females are different, the rate of inbreeding depends mainly on the number of the less numerous sex.

3. Know the history of the broodstock. It is critical to know N_e for each generation, both at your hatchery and before a stock arrives. If a hatchery population is started with only a few individuals a subsequent increase in N_e does not rectify the problems that have occurred. The population may not recover its genetic diversity for many generations after a bottleneck, other than through the introduction of new stock.

4. Cross fish of a different origin (from different lines, strains, or populations) to produce intraspecies hybrids. Intraspecies crossing (crossbreeding) is an effective method for production of hybrids with increased productivity (part 5.4.1). Crossbreeding also is an effective tool to increase genetic diversity and to prevent inbreeding. If two distant, according to their origin, strains are crossed, the coefficient of inbreeding (F) of the resulting progeny will be close to 0.

5. Maintain breeding records and use special schemes of crosses for eliminating the possibility of mating close relatives. One of these schemes (so called 'rotational line crossing') is shown in Figure 5.3.2. You can see in the figure that this scheme of crosses eliminates the possibility of crossing sibs. Lines A, B, and C in parental generation (P) may differ by origin. However, this scheme of crosses may also be used when an initially homogeneous broodstock, presented at a hatchery, is arbitrarily divided into several groups.

5.4. Intraspecies and Interspecies Hybridization

5.4.1. Intraspecies Hybridization or Crossbreeding

Intraspecies hybridization or **crossbreeding** is crossing of animals belonging to the same species but to different strains, lines, breeds or populations.

Crossbreeding may be used:

1. In selection programs

2. For commercial hybridization

In selection programs, crossbreeding is used for so-called synthetic selection, when the final purpose is to combine some useful traits of two groups of fish. Two

Figure 5.3.2. Scheme of rotational line crossing (from Kincaid 1977; reproduced with permission from American Fisheries Society).

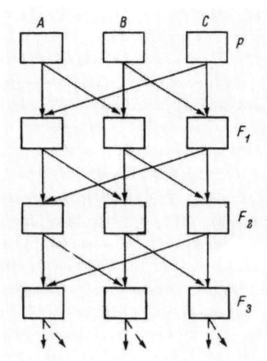

main types of crosses, reproducing and introductory, are used in synthetic selection (Figure 5.4.1). **Reproducing cross** (Figure 5.4.1.A) is used when two groups of fish with valuable traits are crossed and the obtained hybrids are reproduced in subsequent generations. For example, this cross was used for development of a common carp breed "Ropsha" resistant to cold climate in Russia (Kirpichnikov 1981). In this case a domesticated, fast-growing line of common carp was crossed with a tolerant wild-type subspecies of common carp. The purpose of **introductory cross** (Figure 5.4.1.B) is to introduce some useful trait (for example, resistance to disease) from one group of fish (B in Figure 5.4.1.B) to another group of fish (A). In each generation obtained hybrids are backcrossed with group A.

Figure 5.4.1. Scheme of reproducing (A) and introductory (B) crosses used in synthetic selection.

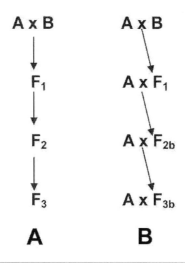

Application of crossbreeding in commercial hybridization is based on the fact that animals, obtained by crossing different lines or strains, frequently have better productive traits as compared to parental forms. This phenomenon is called **heterosis** or hybrid vigor. There are two main theories explaining appearance of heterosis. The "theory of dominance" explains heterosis by suppression of deleterious recessive alleles from one strain in hybrid genomes by dominant alleles from the other strain. The "theory of overdominance" explains the appearance of heterosis by positive action of heterozygosity. It is supposed that heterozygotes (*Aa*) have advantage over both types of homozygotes (*AA* and *aa*). Heterosis in intraspecies crosses is a widespread phenomenon among different animals including fish. Hybrid vigor usually has the maximum effect in the first (F_1) generation. In F_2 and in subsequent generations it diminishes. Heterosis is based on non-additive genetic variance, which results from allele and gene combinations. As indicated in Chapter 3, non-additive genetic variance cannot be kept by selection.

Two-line breeding is frequently used in aquaculture. According to this scheme, two lines are reproduced separately, but only hybrids between these two lines are reared commercially. The scheme of two-line breeding is shown in Figure 5.4.2. For example, production of common carp in Israel is organized on the basis of two-line breeding. Fish hatcheries keep separately two lines of common carp (Dor-70 and Našice). All common carp reared in the country commercially are hybrids between these two lines.

Figure 5.4.2. Scheme of two-line breeding in aquaculture.

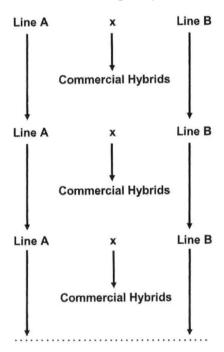

In order to establish two-line commercial breeding, it is preferred to evaluate several crosses (different combinations of 3 or more strains) to find out the best cross.

Increased productivity of interstrain hybrids has repeatedly been described in many aquaculture species such as rainbow trout, channel catfish, grass carp and others. Crossbreeding can be considered as a relatively simple and inexpensive method when compared with selection programs. Crossbreeding is the optimal way to prevent inbreeding depression in commercially raised stocks.

5.4.2. Interspecies Hybridization

Interspecies (or distant) hybridization is the crossing of animals belonging to different species. External fertilization and high fecundity are the factors which make distant hybridization in fish easy. Numerous cases of interspecies hybridization in fish in natural conditions have been described. By means of artificial insemination hundreds of different interspecies hybrids in fish were also obtained.

Interspecies hybridization differs to a great extent from intraspecies hybridization, which was described in the previous part of this chapter. Intraspecies hybrids have in their genomes paternal and maternal haploid chromosome sets of the same species. In the case of interspecies hybridization, the haploid chromosome sets of two different species (and egg cytoplasm of maternal species) are combined. Maternal and paternal chromosome sets may differ with regard to the number of chromosomes, gene composition, amount of DNA content, etc. Therefore, results of distant hybridization may be very different. To systematize data on distant hybridization in fish, it is reasonable to distinguish somatic and reproductive performance of hybrids.

The scheme in Figure 5.4.3 represents possible differences in success of hybridization and somatic performance of interspecies hybrids in fish. You can see in the figure that results of hybridization may be very different. Description of each category (I-IV+) shown in this scheme (Figure 5.4.3) is given below:

I. No fertilization. Table 5.4.1 presents the results of crosses of black crappie females with males of different species. You can see that sperm taken from common carp and blue sucker males were not able to fertilize black crappie eggs (or fertilization rate was extremely low).

II. Hybrid larvae are inviable. As you can see in Table 5.4.1, after insemination

Figure 5.4.3. Success of hybridization in fish and somatic performance of interspecies hybrids.

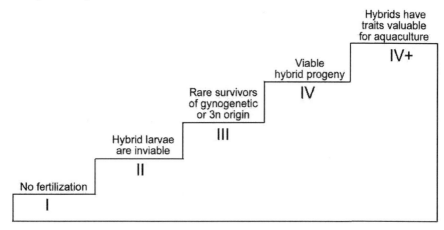

Table 5.4.1. Results of crosses black crappie females with males of different species (from Gomelsky et al. 2000; reproduced with permission from American Fisheries Society).

Cross[a]	Number of eggs	Number of fertilized eggs	Fertili- zation rate (%)	Hatchlings		Normal larvae
				Number	Percent[b]	
♀BC × ♂WB	1,635	966	59.1	900	93.2	0
♀BC × ♂WE	1,641	927	56.5	165	17.8	0
♀BC × ♂CC	1,510	9	0.6	0	0	
♀BC × ♂BS	~1,500	0	0			
♀BC × ♂BC[c]	1,088	924	84.9	910	98.5	828

[a] BC = black crappie, WB = white bass, WE = walleye, CC = common carp, BS = blue sucker.
[b] From the number of fertilized eggs.
[c] Control.

of crappie eggs with sperm of walleye or white bass, fertilization rate was relatively high (55-60%). However, all resulting embryos had various morphological abnormalities and died either before or soon after hatching. Results of these crosses show that combinations of haploid chromosome sets of crappie and walleye or white bass in hybrid genomes were not able to function properly. (Based on the results of these experiments white bass was chosen as the optimal sperm donor for inducing gynogenesis in black crappie.)

III. Rare survivors of gynogenetic or triploid origin. In some distant crosses, very rare (0.1-0.01%) survivors were observed. Analysis showed that these individuals are not normal diploid hybrids (i.e. having two haploid chromosome sets of two species) but are of gynogenetic or triploid origin.

Viable gynogenetic fish are produced when male chromosomes are spontaneously inactivated and the female haploid chromosome set is duplicated by spontaneous suppression of the 2nd meiotic division in the eggs (we will study gynogenesis in Chapter 6). Diploid gynogenetic fish are not hybrids at all; they have two haploid chromosome sets from the maternal species.

Triploids are produced when male chromosomes participate in fertilization but the female chromosome set is doubled by spontaneous suppression of the second meiotic division. Triploid survivors were observed, for example, after crossing females common carp (2n=100) with males of grass carp (2n=48) (Vasilyev et al. 1975). The karyotype of rare survived fish had 124 chromosomes, i.e. it consisted of a diploid set from common carp and a haploid set from grass carp (i.e. $2n_1 + n_2$). (materials on triploidy are given in Chapter 6). All diploid hybrids between these two species having 74 chromosomes (i.e. n_1+n_2) were inviable.

IV. Viable hybrid progeny. This category includes numerous cases when true hybrids, having two haploid chromosome sets from two species, are viable. As a rule, morphologically, hybrids are intermediate between two species. Sometimes, two reciprocal hybrid forms differ; in these cases hybrids are more similar to the maternal species.

IV+. Hybrids have traits valuable for aquaculture. This category is a part of the previous category and contains interspecies hybrids having traits valuable for aquaculture. A brief description of several hybrids that are used in aquaculture is given below:

- **Hybrids striped bass.** Hybrids between striped bass (*Morone saxatilis*) and white bass (*Morone chrysops*) are popular aquaculture objects. Both reciprocal hybrids, palmetto bass (female striped bass x male white bass) and sunshine bass (female white bass x male striped bass) have a fast growth rate and are commercially raised in the United States and some other countries. However, sunshine bass are more popular since females of white bass are more available and easier to spawn.

- **Hybrid catfish (female channel catfish, *Ictalurus punctatus* x male blue catfish, *Ictalurus furcatus*).** Compared to channel catfish, these hybrids exhibit superior characteristics for many traits such as growth rate, tolerance of low oxygen, and resistance to diseases. Rearing of these hybrids on a large scale is restricted by the difficulty of obtaining fish for stocking since breeders of channel catfish and blue catfish do not spawn naturally. Therefore hybrids may be obtained only by artificial spawning.

- **Hybrids between tilapia species.** The initial advantage for producing hybrids in tilapias is all-maleness of progenies obtained in several hybrid combinations. Sometimes, interspecies tilapia hybrids have good growth rates, also. (Taiwanese red tilapia, which had hybrid origin, was described in Chapter 1.).

- **Hybrid sunfish (female green sunfish, *Lepomis cyanellus* x male bluegill, *Lepomis macrochirus*).** These hybrids have a good growth rate and they are not as fertile as the parental species since the sex ratio is skewed towards males (up to 90%).

- **Hybrid between female walleye (*Sander vitreus*) and male sauger (*S. canadensis*)** (so called "saugeye"). These hybrids have faster growth rate and better survival in ponds and impoundments than parental species. In several US states saugeye is popular game fish and widely stocked. However, these hybrids are fertile; therefore their stocking may result to genetic contamination of parental species.

- **Hybrid between sturgeon species, beluga (*Husu huso*) and sterlet (*Acipenser ruthenus*).** These hybrids (called "bester") have a fast growth rate from beluga and the ability to be raised in still-water ponds from sterlet. Originally bester have been obtained in Russia; currently these hybrids are raised in several European and Asian countries.

We considered above somatic performance of interspecies hybrids. However, reproductive performance of hybrids is a subject of special consideration. Figure 5.4.4 presents the possible influence of interspecies hybridization on reproductive ability of fish. You can see in the figure that reproductive performance of distant hybrids may be very different. A description of each category (I-IV) shown in scheme in Figure 5.4.4 is given below:

I. Hybrid males and females are genetically sterile. Sterility is regarded as a common trait for interspecies hybrids in animals. Somatically hybrids may be perfect, but reproductively sterile. What is the reason for this drastic difference? As we know, somatic cells divide by mitosis. During mitosis there is no conjugation of homologous chromosomes and non-compatibility of maternal and paternal chromosome sets in hybrids does not influence somatic cell divisions. The situation is changed when the meiotic process in sex cells begins. Incompatibility of parental chromosome sets in hybrids may result in numerous disturbances during meiotic process. Abnormal conjugation and non-disjunction of chromosomes may result in complete blockage of meiosis and gametogenesis. 'Genetic sterility' means that hybrids are not capable of producing viable progeny. In addition to genetic sterility, morphologically, hybrid females and males can have reduced size gonads.

Sterility of distant hybrids in fish is a fairly common phenomenon (the same as for other animals). Sometimes sterility is a useful trait. For example, hybrids between brown trout (*Salmo trutta*) and brook trout (*Salvelinus fontinalis*) (so called "tiger trout") are sterile and therefore can be stocked into natural water bodies where reproduction is not desirable.

Figure 5.4.4. Reproductive performance of interspecies hybrids in fish.

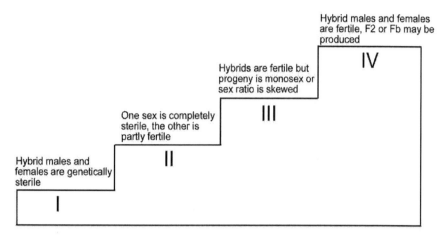

II. One sex is completely sterile, the other is partly fertile. In this category of hybrids, males are sterile but some females managed to restore fertility by transformation of meiotic process. In early oogenesis additional reduplication of chromosomes occurs and, as a result, females produce eggs with somatic number (2n) of chromosomes (i.e. without reduction in the number of chromosomes). This phenomenon was described in several fish hybrids belonging to very different systematic groups. Some of these hybrids are listed below:

- Silver crucian carp x common carp (family *Cyprinidae*) (Cherfas et al. 1994);
- Atlantic salmon x brown trout (*Salmonidae*) (Johnson and Wright 1986; Galbreath and Thorgaard 1995)
- pumpkinseed x green sunfish (*Centrarchidae*) (Dawley et al. 1985)

Because of egg diploidy, crossing of these hybrid females with males of either parent species results in production of triploid progenies without application of any treatments (shocks) to early embryos.

The examples listed above show that interspecies hybridization can result in transformation of meiosis during oogenesis and trigger ability of females to produce unreduced diploid eggs. As noted in Chapter 1, non-reduction of chromosome number during oogenesis is one of cytological features of natural gynogenesis. In this connection it seems logical that many gynogenetic forms in fish have hybrid origin (see 1.5). There is a hypothesis that interspecies hybridization was the first step towards appearance of any gynogenetic form. (The next and final step should be the ability to inactivate male chromosomes).

III. Hybrids are fertile but progeny is monosex or sex ratio is skewed. Hybridization may cause disturbances in mechanisms of sex determination. As mentioned above, crossing of some tilapia species results in appearance of all-male progenies. But more frequently a shift in sex ratio is observed. For example, hybrid progenies between green sunfish and bluegill consist of 80-90% males.

IV. Hybrid males and females are fertile; F_2 or backcross hybrids (F_b) may be obtained. Fertility of F_1 hybrids does not mean that gametes, which they produce, are genetically balanced. Even if haploid chromosome sets of parental species are similar and relatively normal conjugation occurs, due to independent segregation of chromosomes during meiosis, gametes, formed by F_1 hybrids, carry different numbers and compositions of chromosomes of the parental species. Therefore F_2 hybrids have much higher variability, if compared with F_1 hybrids, and significantly lower performance. It was shown for several distant hybrids in fish such as hybrids striped bass, hybrids channel catfish x blue catfish, and hybrids beluga x sterlet.

In conclusion, it should be noted that interspecies hybridization should always be under control. Uncontrolled hybridization can result in genetic contamination of parental species (both natural populations and artificially reproduced stocks). For example, uncontrolled hybridization of tilapias resulted in a situation when it is hard to find pure species on many commercial fish farms. As demonstrated in Chapter 4, there are cases when morphologically fish breeders are identified as pure species but application of DNA markers shows that these fish are hybrids.

Chapter 6. Chromosome Set Manipulation and Sex Control

6.1. Induced Gynogenesis

6.1.1. Methods of Production of Gynogenetic Fish. Types of Induced Gynogenesis

In Chapter 1 natural gynogenesis in fish was described, a mode of reproduction that was revealed in several all-female forms. **Gynogenesis** is a type of reproduction where the male chromosomes are inactivated after insemination and embryonic development is controlled only by female chromosomes.

Gynogenetic progenies may be obtained in species reproducing by the normal sexual mode. The main cytogenetic features of gynogenesis, inactivation of male chromosomes and prevention of female chromosome set reduction, can be achieved by special experimental treatments.

Inactivation of male chromosomes is achieved by irradiation of sperm. X-rays, gamma rays, or ultraviolet (UV) irradiation may be used for this purpose. This method of chromosome inactivation is based on different sensibilities of chromosomes and cytoplasmatic structures of spermatozoa. At high dosage of irradiation, chromosomes are inactivated but spermatozoa maintain ability for movement and insemination of eggs.

The optimal dosage of irradiation is determined according to the **'Hertwig effect'** and haploidy of obtained larvae. The 'Hertwig effect' is the paradoxical improvement of embryo viability along with increase of irradiation dose. This phenomenon was named after German scientist Oscar Hertwig, who described it in 1911.

Figure 6.1.1 presents the Hertwig effect in hatchability of larvae observed after insemination of black crappie eggs with irradiated white bass sperm in study performed by Gomelsky et al. (2000). You can see that percentage of hatched larvae decreased for doses from 0 to 100 J/m^2 because of damage to male chromosomes and appearance of embryos with heavy morphological abnormalities. However, with further increase in dose, the hatching rate paradoxically increased. This increase is explained by the fact that doses higher than 100 j/m^2 began to induce complete inactivation of male chromosomes and embryo development switched to haploid gynogenesis. Haploids are inviable, but they perished usually after hatching. At dosages 1000 j/m^2 and higher the percentage of hatched larvae was stabilized at 80-85%. It means that inactivation of male chromosomes was observed in all embryos.

Haploidy of gynogenetic larvae of black crappie, obtained at high dosage of sperm irradiation, was confirmed by determination of DNA content using flow cytometric analysis as seen in Figure 6.1.2.

Figure 6.1.1. 'Hertwig effect' in hatchability after insemination of black crappie eggs with UV-irradiated white bass sperm (from Gomelsky et al. 2000, reproduced with permission from American Fisheries Society).

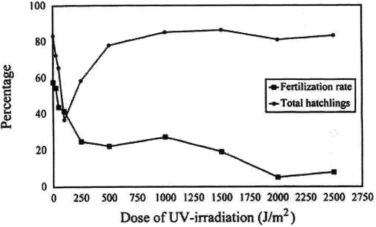

You can see in Figure 6.1.2 that haploid larvae (B) have one-half of the DNA content of normal diploid larvae (A). Haploidy may also be revealed by counting of chromosomes.

Figure 6.1.3 presents a scheme of induced gynogenesis in fish. Figure 6.1.3.A shows production of haploids, i.e. haploid gynogenesis. Haploids in fish are morphologically abnormal and die before or soon after hatching. In order to obtained viable gynogenetic fish, haploid female chromosome set should be doubled. The female chromosome set may be doubled two ways: either by suppression of the

Figure 6.1.2. DNA content in diploid (A) and haploid (B) larvae of black crappie (from Gomelsky et al. 2000; reproduced with permission from American Fisheries Society).

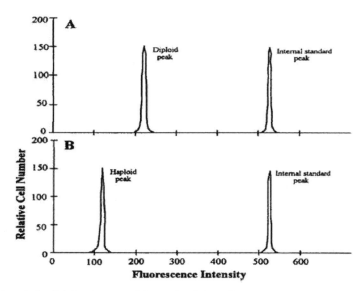

second meiotic division in eggs (Figure 6.1.3.B – diploid meiotic gynogenesis) or by suppression of the first mitotic division in haploid embryos (Figure 6.1.3.C – diploid mitotic gynogenesis).

For artificial suppression of meiotic or mitotic divisions, strong physical treatments (shocks) are applied to the embryos. Most commonly used treatments are low and high temperatures (cold and heat shocks) or hydrostatic pressure. For suppression, physical treatment is applied at anaphase of the corresponding division. Under the influence of treatment, a spindle of the division is destroyed and, as a result, division stops and its daughter products are fused. When the second meiotic division is suppressed, the 2nd polar body is not extruded (see Figure 6.1.3.B); it is fused with the haploid female pronucleus. During suppression of the first mitotic division in haploid embryos, two haploid nuclei are united to form a

Figure 6.1.3. Scheme of induced gynogenesis in fish. A - haploid gynogenesis; B - diploid meiotic gynogenesis; C - diploid mitotic gynogenesis; p.b. – polar body; inactivated male chromosomes are depicted as small dots.

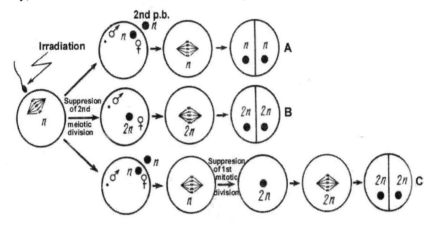

diploid nucleus (see Figure 6.1.3.C). After the next mitotic cycle two diploid blastomeres are formed.

The existence of two mechanisms of induced diploid gynogenesis is clearly visible in results of experiment performed by Gomelsky with colleagues on induced gynogenesis in common carp, which are shown in the Figure 6.1.4. This figure presents the yield of diploid gynogenetic larvae depending on the timing of heat shock application; the shock temperature was 40° C; shock duration was 3 min. You can see in the figure that yield of diploid larvae had two peaks. The first peak, observed after application of heat shock at 3 min after insemination (so called "early shock"), was caused by suppression of the second meiotic division in eggs (meiotic gynogenesis), while the second peak, observed after application of shock at about one hour after insemination ("late shock"), was caused by suppression of the first mitotic division in haploid embryos (mitotic gynogenesis).

Techniques for production of meiotic and mitotic gynogenetic progenies were elaborated for many aquaculture species. For each species the optimal parameters of sperm irradiation (dosage) and shocks (temperature or pressure level, duration, timing) were determined.

Figure 6.1.4. Yield of diploid gynogenetic larvae in common carp by application of heat shock at different time after insemination.

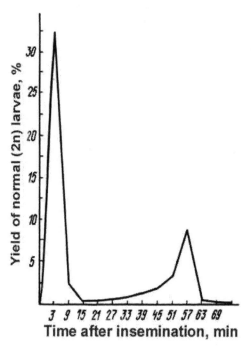

Gynogenetic origin of fish produced in experiments may be confirmed either by application of genetic markers or by using a sperm from other fish species for stimulation of gynogenetic development. When morphological genetic markers are used for this purpose, eggs taken from female with recessive trait are inseminated with irradiated sperm taken from male with dominant trait. If chromosomes of spermatozoa are successfully inactivated by irradiation, offspring will have maternal recessive trait. This will indicate to their gynogenetic origin. Example: Eggs from mirror common carp (*ss*) are inseminated with irradiated sperm from scaled male (*SS*). Gynogenetic offspring should be mirror (*ss*). Appearance of scaled individuals (*Ss*) in progeny will indicate to improper irradiation of sperm.

Currently DNA markers are widely used for confirmation of gynogenetic origin of offspring. It is convenient to use for this purpose RAPD markers, which are characterized by dominance between alleles (see part 4.3.4). Gynogenetic offspring should not have specific paternal RAPD bands. Application of codominant microsatellites as markers for proving of exclusion of paternal inheritance is based on large number of alleles per locus (the same as parentage determination, see part 4.4.4). Gynogenetic offspring should not have microsatellite alleles which male, whose sperm was irradiated, possessed.

When sperm from another fish species is used for stimulation of gynogenetic development, the optimal condition is inviability of normal diploid hybrids between two species. Even if spermatozoa are not irradiated properly for any reason, the resulting hybrids will not survive. For example, as was mentioned above, for induction of diploid gynogenesis in black crappie, irradiated sperm of white bass was used. As shown in part 5.4.2, hybrids between black crappie and white bass are inviable.

6.1.2. Properties of Gynogenetic Fish

Induced gynogenesis results in increased homozygosity. Therefore, it may be regarded as a form of inbreeding. The rate of homozygosity increase depends on the type of gynogenesis.

In the case of **meiotic** gynogenesis both types of homozygotes (*AA* and *aa*) and heterozygotes (*Aa*) may be found in gynogenetic progenies obtained from female heterozygous for some gene (genotype *Aa*). Genotype of gynogenetic fish depends on allele recombination in prophase I in oocytes (Figure 6.1.5). If in chromosome region between gene and centromere there was no crossing over, only homozygous fish appear (*AA* and *aa*) (Figure 6.1.5.A). Crossing over results in appearance of heterozygotes; mechanism of their appearance is shown in Figure 6.1.5.B. It should be noted that crossing over is usually recorded and analyzed when several (at least two) genes located in one chromosome. In case of meiotic gynogenesis in fish we can observe result of crossing over based on behavior of one gene. Description of similar phenomenon of crossing over between gene and centromere can be found in textbooks on general genetics in chapters on linkage analysis in fungus *Neurospora*.

Figure 6.1.5. Scheme of allele recombination in case of inducing meiotic diploid gynogenesis from heterozygous female (*Aa*); a - primary oocytes, b - secondary oocytes, c - gynogenetic diploids.

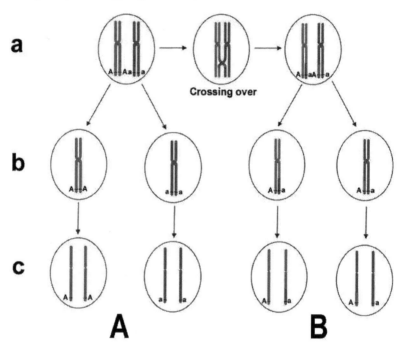

Frequency of crossing over depends on the distance between a given gene and centromere and may differ to a great extent. Therefore the frequency of heterozygotes in meiotic gynogenetic progenies obtained from heterozygous females varies for different genes from 0 to 100%. The average frequency of heterozygotes for gene is about 50% (although there is some variability between different species). On this basis, coefficient of inbreeding (F) for one generation of meiotic gynogenesis is close to 0.50 (the value of F which is typical for self-fertilization, see part 5.3).

Mitotic gynogenesis is an extreme form of inbreeding. Coefficient of inbreeding for one generation of mitotic gynogenesis is 1.0 (i.e. homozygosity for all genes). Mitotic gynogenetic progenies obtained from female heterozygous for some gene (genotype *Aa*) consist of two clases of homozygotes (*AA* and *aa*) and no heterozygotes are present. Observed homozygosity for all genes in mitotic gynogenetic progenies is explained by the fact that in this case homologous chromosomes are the products of simple mitotic reduplication in haploid embryos. Therefore fish obtained by mitotic gynogenesis are sometimes called "double haploids".

Using mitotic gynogenesis it is possible to produce clones in fish, i.e. groups of genetically identical individuals. Clones in fish may be produced by two methods:

1. From females obtained by means of mitotic gynogenesis, the second consecutive gynogenetic generation (meiotic or mitotic) is obtained. (Usually meiotic gynogenesis is used since at time of "early shock" embryos are less sensitive to physical treatments than at time of "late shock".) Resulting individual progenies will be clones and each fish in the progeny will have the maternal genotype and, the same as mother, will be homozygous for all genes. Therefore, clones produced by this method are called "homozygous". In Figure 4.3.3 (Chapter 4) you can see that fingerprints of mother and fish from homozygous clone are identical.

2. Females, obtained by mitotic gynogenesis, are crossed with sex-reversed males of the same origin (materials on hormonal sex reversal are given in part 6.4). Resulting clones are called "heterozygous" since heterozygosity for some genes can result from difference in allele composition between homozygous for all genes fish breeders (i.e. females and sex-reversed males) of mitotic gynogenetic origin.

For the first time clones in fish by application of induced gynogenesis were obtained in a model species zebrafish (*Danio rerio*) by Streisinger et al. (1981). Up to the present, clones have been produced in several aquaculture species such as common carp, rainbow trout, tilapia, red sea bream and some others.

In the case of female homogamety (females - XX, males - XY) gynogenetic progenies are all-female. Nevertheless, this method is not usually used for direct sex regulation in fish for these reasons:

- Possible signs of inbreeding depression observed in gynogenetic progenies;
- Relative complexity.

However, all-female gynogenetic progenies are frequently used for initial experiments on hormonal sex reversal (see 6.4).

In the case of female heterogamety (females - WZ; males - ZZ,) gynogenetic progenies consist of both females and males; the percentage of males may be different in different species. As indicated in Chapter 1, according to sex composition in gynogenetic progenies the mechanism of sex determination (i.e. either female or male heterogamety) in a given species may be identified.

6.2. Induced Androgenesis

Androgenesis (from Greek *andro* - male, and *genesis* - origin) is embryo development under the control of only paternal chromosomes.

In contrast to gynogenesis, there is no natural androgenesis as a mode of reproduction in fish. However, androgenetic development may be induced in species reproducing by means of the normal sexual mode. In this case, inactivation of female chromosomes is achieved by irradiation of eggs. The scheme of induced androgenesis in fish is presented in the Figure 6.2.1. After insemination of eggs with genetically inactivated chromosomes by intact spermatozoa, androgenetic haploids are produced (Figure 6.2.1, A). To produce viable diploid androgenetic fish one needs to double the paternal haploid chromosome set. It may be achieved by suppression of the first mitotic division in haploid embryos (Figure 6.2.1, B). By this method androgenetic progenies were obtained in several aquaculture species such as common carp, rainbow trout and tilapia.

Figure 6.2.1. Scheme of induced androgenesis in fish. A - haploid androgenesis; B - diploid androgenesis; inactivated female chromosomes are depicted as small dots.

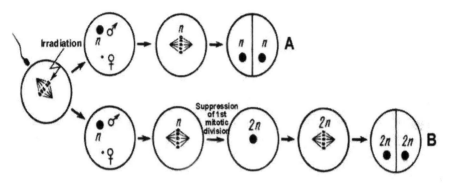

For confirmation of androgenetic origin of obtained progeny, eggs taken from female with dominate trait should be irradiated and then be inseminated with intact sperm taken from male with recessive trait. If chromosomes of eggs are successfully inactivated by irradiation, offspring will have paternal recessive trait. This will indicate to their androgenetic origin. The same as fish obtained by mitotic gynogenesis, androgenetic diploids are homozygous for all genes. Therefore, like mitotic gynogenesis, androgenesis may be used for production of clones.

In species with male heterogamety (males - XY, females - XX) androgenetic progenies consist of females XX and males YY. In species with female heterogamety (males - ZZ, females - ZW) only males ZZ are present in androgenetic progenies.

The method of induced androgenesis is of special interest for preservation of genetic variability and restoration of rare and endangered species. Presently, techniques of sperm cryopreservation have been elaborated for many fish species. Obtaining diploid androgenetic progenies using cryopreserved sperm creates opportunities to restore genomes.

6.3. Induced Polyploidy

6.3.1. Production of Triploid Fish by Suppression of 2nd Meiotic Division in Eggs

Induced polyploidy is artificial production of individuals with an increased number of haploid chromosome sets. In aquaculture and fisheries this method is primarily used for production of triploid fish, i.e. fish whose karyotypes contain three haploid chromosome sets.

Triploid fish are genetically sterile, i.e. they are not capable of producing viable progeny. Also triploids are characterized by complete or partial reduction of gonads. Abnormalities in the development of reproductive systems of triploids are caused by the presence of a third, "additional" haploid chromosome set, which disturbs the normal process of conjugation and disjunction of homologous chromosomes.

Genetic sterility and reduction in gonad development of triploid fish provides two possible benefits for their production in aquaculture and fisheries. Triploid fish may be produced:

- To prevent fish reproduction in ponds, reservoirs, or natural waters.
- To obtain a better growth rate as compared with normal diploids.

There are several methods for producing triploid fish. The most simple and better developed method is based on the suppression of the 2nd meiotic division in eggs after insemination by intact (non-irradiated) spermatozoa. The scheme of this method is given in the Figure 6.3.1. As you can see in the figure, after suppression of the second meiotic division, a diploid female pronucleus fuses with a haploid male pronucleus and a triploid zygote is formed. For production of triploid fish, the same physical treatments (shocks) of eggs are used as for induction of meiotic gynogenesis. Shock application should provide a high frequency of triploids, but at the same time does not decrease drastically embryo survival. Techniques for production of triploid progenies by suppression of the 2nd meiotic division in eggs have been elaborated for many aquaculture species. Results of some studies on induction of triploidy by suppression of 2nd meiotic division in eggs in several aquaculture species are presented in Table 6.3.1. You can see in the table that in many cases applied shock results in a high percentage of triploids in progenies (up to 100%) along with relatively high (more than 50%) larvae viability.

Figure 6.3.1. Scheme of production of triploids by suppression of 2nd meiotic division in eggs.

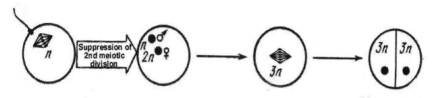

Table 6.3.1. Results of some studies on induction of triploidy by suppression of 2nd meiotic division in eggs.

Species	Type of treatment	Parameters of treatment*	% of triploids	Yield of viable larvae, %	Reference
Common carp	Heat shock	39-42°C, 5-6 min a.i., 1-3 min	85-100	Up to 60	Recoubratsky et al. 1992
Grass carp	Hydrostatic pressure	560 kg/cm², 4 min a.i., 1-2 min	Close to 100	40-55	Cassani and Caton 1986
Rainbow trout	Heat shock	26-28°C, 40 min a.i., 10 min	90-100	50-57	Solar et al. 1984
	Hydrostatic pressure	493 kg/cm², 40 min a.i., 3-4 min	100	70-85	Chourrout 1984
Channel catfish	Cold shock	5 °C, 5 min a.i., 60 min	100	80	Wolters et al. 1981

* Temperature or pressure level, timing in minutes after insemination (a.i.) and duration.

6.3.2. Properties of Triploid Fish

Reproductive performance of triploids. In the case of male heterogamety (XX/XY) triploid progenies consist of males XXY and females XXX with sex ratio 1:1. In the case of female heterogamety (WZ/ZZ) triploid progenies consist of females WZZ and WWZ, and males ZZZ. Sex ratio in triploid progenies in species with female heterogamety may be different; usually the prevalence of females (60-80%) is observed.

Triploid females and males differ significantly with regard to the rate of gonad development. In females, disturbances in meiosis stop oogenesis at early stages. Therefore, ovaries in triploid females have usually a very small size. In triploid males, disturbances in the process of meiosis are also observed; spermatogenesis usually stops during meiotic divisions. However, suppression of spermatogenesis in triploid males does not prevent active mitotic proliferation and entering into meiosis of new generations of spermatogonia. As a result, testes in triploids may reach relatively large size. For example, Peruzzi et al. (2004) showed that in triploid females of sea bass (*Dicentrarchus labrax*) a gonadosomatic index (GSI, percentage of gonad weight to body weight) was about 90 times lower than in diploid females while GSI in triploid males was only twice lower than in diploid males. This difference in development of ovaries and testes in triploid and diploid sea bass is clearly visible in Figure 6.3.2.

Figure 6.3.2. Gonadal condition in triploid (3n) and diploid (2n) sea bass; a – ovaries, b – testes (from Peruzzi et al. 2004, Aquaculture 230:41-64; Copyright 2004, reproduced with permission from Elsevier).

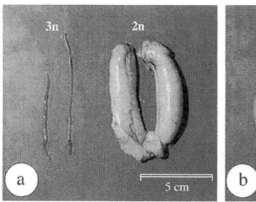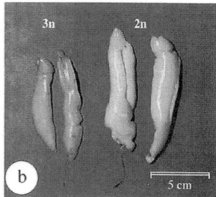

Sometimes triploid males produce small amounts of sperm. Insemination of eggs with sperm from triploid males results in the appearance of inviable embryos since spermatozoa, originated from atypical triploid meiosis, carry unbalanced chromosome numbers. Studies conducted in several aquaculture species confirmed genetic sterility of triploid fish and the possibility of stocking these fish in water bodies for preventing of reproduction.

Stocking of triploid grass carp (*Ctenopharyngodon idella*, family *Cyprinidae*) in the United States is an example of successful application of triploidy for this purpose. According to regulations accepted in about 30 US states, only sterile triploid grass carp may be stocked for aquatic weed control. Triploid grass carp for stocking are produced by commercial fish hatcheries. Commercial batches of triploid grass carp for stocking were produced for the first time in 1983. The triploidy of each produced fish is checked with a Coulter Counter (see below); this procedure is under supervision of US Fish and Wildlife Service.

In several US states (for example, Washington and California) sterile triploid rainbow trout or other salmons are produced and stocked in water bodies as sport fish.

Somatic performance of triploids. Polyploidy does not increase directly the body size in fish. It is known that at the beginning of maturation the rate of somatic growth in fish is significantly decreased due to expending of nutrients and energy to development of gonads. It was supposed that such retardation in growth rate would not be observed in triploids due to nondevelopment of their gonads. This would give an opportunity to get additional production without additional expenses.

Many studies were performed to compare the growth rate of diploid and triploid fish. The obtained results are ambiguous: some studies reported a growth advantage of triploids, whereas others found similar or even reduced growth of triploids if compared with diploids. Even if some advantage in the growth rate of triploids were revealed the following factors prevent commercial production of triploid progenies:

- Relatively complex procedures of egg treatments should be used each time for production of triploid progenies. Results of trials for production of triploids are not consistent.

- Treatments of eggs with temperature or pressure shocks frequently results in decreased survival especially during earlier stages of rearing.
- Gonad development reduction in triploid males is not so profound. Therefore it was suggested to produce all-female triploid progenies by joint application of induced polyploidy and genetic sex regulation (part 6.4).

Induced triploidy and hybridization. Induced polyploidy may be combined with distant hybridization. Polyploids having haploid chromosome sets of one species are called **autopolyploids**. Polyploids having haploid chromosome sets of different species are called **allopolyploids**.

Allotriploids in fish may be obtained by suppression of the 2nd meiotic division in eggs after insemination with sperm of another species. In this case the resulting fish have two haploid chromosome sets of the maternal species and one haploid set of paternal species. Sometimes allotriploid hybrids have valuable traits. For example, allotriploid hybrids between rainbow trout and coho salmon have higher viability than corresponding diploid hybrids and increased tolerance to diseases. Triploid hybrids between female walleye and male sauger are sterile (in contrast to fertile diploid hybrids of these species) and therefore may be stocked in natural water bodies without danger of genetic contamination with parental species.

Methods of fish ploidy determination. Methods of ploidy determination may be direct and indirect. Direct methods are based on the analysis of karyotypes or content of DNA. Content of DNA is determined by means of flow cytometric analysis. Triploid fish have three haploid sets in karyotypes and 1.5 times more DNA per nucleus than diploids.

Indirect methods of ploidy determination are based on the fact that polyploidy in fish (as well as in other animals) increases proportionally the size of cells and nuclei. Therefore, in triploid fish the volume of cells and nuclei is 1.5 times larger than in diploid fish. Usually, ploidy of fish is determined by measurement of erythrocyte nuclei on stained blood smears. In Figure 6.3.3 microphotographs of erythrocytes of diploid and triploid common carp are shown. You can see that the size of cells in the diploids and triploids visually differ to a great extent. The calculated ratio of cell or nuclei squares in diploid and triploid fish is 1.0 to 1.31.

Figure 6.3.3. Microphotographs of erythrocytes of diploid (A) and triploid (B) common carp.

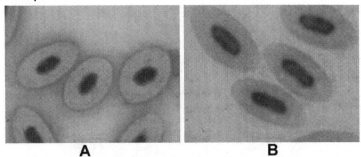

A   **B**

A special laboratory device called a Coulter Counter permits automation of the process of ploidy determination in fish. This device is able to record electronically the size of passing particles. Wattendorf (1986) has described technique of using Coulter Counter for determination of ploidy in grass carp on basis of difference in size of erythrocytes between diploid and triploid fish. Application of this device provides a reliable and rapid identification of ploidy for large quantities of fish. Currently it is a standard procedure used in private fish hatcheries for verification of ploidy in grass carp (Figure 6.3.4).

6.3.3. Production of Triploids by Crossing of Tetraploids

The second possible method for production of triploid fish consists of two consecutive steps:

1. Production of tetraploid fish;
2. Crossing of tetraploid breeders with normal diploid fish.

Tetraploid fish are produced by suppression of the first mitotic division in normal diploid embryos. Physical treatments (shocks) are applied at anaphase of first mitotic (cleavage) division (the same as for induction of mitotic gynogenesis). The scheme of production of tetraploid fish is given in the Figure 6.3.5.

Figure 6.3.4. Verification of grass carp ploidy at private fish hatchery, J.M. Malone and Son, Inc. (Lonoke, Arkansas, USA); A – taking blood sample from anesthetized fish; B – checking blood sample with Coulter Counter (photos by the author).

Figure 6.3.5. Scheme of production of tetraploid fish by suppression of first mitotic division in diploid embryos; p.b. – polar body.

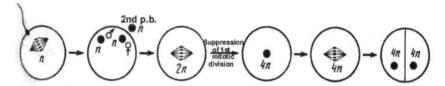

In general, meiosis in tetraploids is normal. During meiotic conjugation in tetraploids, four homologous chromosomes form two pairs. As a result of two meiotic divisions a reduction of chromosome number occurs and diploid (2n) gametes are produced. Due to diploidy of gametes produced by tetraploids it is possible to obtain triploids by crossing them with normal diploids:

$$\male 4n \times \female 2n \to 3n\ \female\female \text{ and } \male\male$$
$$\female 4n \times \male 2n \to 3n\ \female\female \text{ and } \male\male$$

In several fish species such as grass carp, common carp, and walleye tetraploids obtained by suppression of the 1st mitotic division in diploid embryos were inviable and died at the larval stage. Viable tetraploids were produced in other fish such as rainbow trout, Atlantic salmon, tilapia, and yellow perch. Tetraploid had a slow growth rate and decreased survival as compared with diploids and triploids.

Mature tetraploids of rainbow trout, after crossing with normal diploid fish, really produced triploid progenies (Chourroutt et al. 1986a). This method of triploid production is convenient since there is no need to apply treatment each time. However, this method requires production and keeping tetraploid broodstock which can have decreased viability, slow growth rate and higher susceptibility to diseases.

The interesting results of experiments performed by Japanese scientists on transplantation of gonia have been published recently (Okutsu et al. 2007). Spermatogonia isolated from testes of normal (diploid) rainbow trout adult males were injected into body cavity of newly hatched triploid larvae of masu salmon. Injected spermatogonia were incorporated into embryonic gonads of triploid salmon larvae and started normal gametogemesis. As a result, triploid salmons produced rainbow trout gametes (spermatozoa or eggs). Crossing of these two fish resulted in production of rainbow trout offspring. This method does not make triploid fish really fertile; triploid sterile fish are used as host (carrier) for proliferation and development of sex cell from diploid donor. This method can be called "cell engineering".

6.4. Hormonal Sex Reversal and Genetic Sex Regulation

6.4.1. Hormonal Sex Reversal in Fish

Artificial sex regulation is the production of individuals of a desired sex. In fish, females and males sometimes have different economic value. Also rearing or stocking of fish of only one sex gives opportunity to prevent uncontrolled reproduction. Methods of sex regulation in fish are connected mainly with elaboration and application of **hormonal sex reversal** (or inversion).

Hormonal sex reversal is a change from the normal process of sex differentiation under influence of steroid sex hormones so that genotypic females develop testes or genotypic males develop ovaries. Sex reversal changes only fish phenotype, but genotypic formula of sex chromosomes remains the same.

As was noted in Chapter 1, direction of sex differentiation is determined by the action of sex-determining genes. However, the process of sex differentiation is realized through a successive chain of hormonal regulations. On this basis hormonal sex reversal in fish is possible.

The first successful experiments on hormonal sex reversal in fish were performed by the Japanese scientist T. Yamamoto in the 1950-1960s in the model species medaka (*Oryzias latipes*) (review of these studies is given in Yamamoto 1969). By feeding young fish with feed containing female or male steroid hormones he managed to induce sex reversal in genotypic males or females, respectively. For these experiments, a line of medaka with a dominant gene for red color *R* in the Y chromosome was used. Therefore, in this line males (XY) were always red, but females (XX) were white. Yamamoto has fed fry with feed containing female steroid hormone estrone and produced all-female progeny. According to fish color it was easy to distinguish sex-reversed red females (XY) from normal (XX) females. Similarly, when fry were fed with feed containing male steroid hormone methyltestosterone, resulting all-male progeny consisted of sex-reversed white males (XX) and normal red males (XY).

In further experiments Yamamoto crossed sex-reversed fish normal breeders. Sex reversed males XX, when crossed with normal females XX, gave all-female progenies:

$$♂_{rev.}\text{XX} \times ♀\text{XX} \rightarrow 100\% ♀♀\text{XX}$$

Sex-reversed females XY, when crossed with normal males XY, gave mix-sexed progenies with predominance of males:

$$♀_{rev.}\text{XY} \times ♂\text{XY} \rightarrow 25\% ♀♀\text{XX} : 50\% ♂♂\text{XY} : 25\% ♂♂\text{YY}$$

Results of Yamamoto's experiments were a stimulus for studies on hormonal sex reversal conducted in aquaculture and fisheries important fish species. Up to the present, techniques for hormonal sex reversal have been elaborated for many fish. The general rule is that sex reversal may be achieved when hormonal treatment is applied during the period of sex differentiation. Adequate dosage of hormone has to be used.

Tables 6.4.1 and 6.4.2 present techniques of androgen and estrogen treatment, respectively, which have been applied in some studies on hormonal sex reversal in fish. As you can see in these tables, the most common method of hormone (androgen or estrogen) administration to fish is by feeding it with an artificial diet. Other methods of hormone administration, such as immersion in hormone solution or implantation of capsules with hormone, are also used (see Tables 6.4.1 and 6.4.2). These methods are applied in cases when embryos and early larvae should be treated with hormones (chinook, coho and masu salmons) or fish are raised in ponds and can consume natural food (grass carp).

In the case of female homogamety (XX/XY) it is convenient to use gynogenetic progenies consisting of only genotypic females (XX) as material for initial experiments on sex reversal under androgen influence. If normal progenies consisting of genotypic males and females are used, it needs to perform test crosses of males from androgen-treated groups with normal females (XX) in order to

Table 6.4.1. Techniques of fish treatment with androgen (methyltestosterone) for inducing hormonal sex reversal in genotypic females.

Fish species	Method of hormone administration and duration of treatment	Dose of hormone	Reference
Rainbow trout	With dry diet from transition to active feeding during 90 days	3 mg per kg of diet	Johnstone et al. 1978
Chinook salmon	2-hour immersions of larvae in hormone solution 4 and 11 days after hatching (a.h.); from 47 days a.h. with diet for 3-9 weeks	400 µg/L 3-9 mg/kg	Hunter et al. 1983
Common carp	With dry diet to 27-40-day-old fry during 40 days	100 mg/kg	Gomelsky et al. 1994
Grass carp,	Implantation of capsules with slowly releasing hormone into body cavity to 55-day-old fish	12 mg per capsule	Jensen et al. 1983
Crappie	With dry diet from 35 day after hatching during 40 days	30 mg/kg	Cuevas-Uribe et al. 2009
Nile tilapia	With dry diet during 25 or 35 days from the transition to external feeding	60 mg/kg	Tayamen and Shelton 1978

Table 6.4.2. Techniques of fish treatment with estrogen (estradiol) for inducing hormonal sex reversal in genotypic males.

Fish species	Method of hormone administration and duration of treatment	Dose of hormone	References
Rainbow trout	With dry diet from transition to active feeding during 30 days	20 mg/kg	Johnstone et al. 1978
Coho salmon	Repeated 2-hour immersions embryos and larvae to hormonal solution, after swim-up with dry diet during 10 weeks.	50-400 µg/L 10 mg/kg	Goetz et al. 1979
Masu salmon	Keeping larvae in hormonal solution from 5 days after hatching during 18 days.	5-25 µg/L	Nakamura 1981
Yellow perch	With dry diet to 20-35-mm-length fry during 84 days	15-120 mg/kg	Malison et al. 1986
Mozambique tilapia	With dry diet from 6 to 25 days after hatching	50 mg/kg	Nakamura and Takahashi 1973

distinguish sex-reversed males XX and normal males XY. Sex-reversed males XX produce all-female progenies while normal males XY produce normal mixed-sex progenies. As shown in Chapter 4, DNA markers can also be used for determination of formula of sex chromosomes in fish from androgen-treated groups.

As a rule, hormonal sex reversal in fish is complete and functional, i.e. reversed fish function during their life span as normal breeders of opposite sex. For example, according to data by Gomelsky (1985) reversed males XX of common carp did not differ significantly from normal males XY with regard to the amounts of sperm (stripped from fish after hormonal injection) and concentrations of spermatozoa. However, reversed males of rainbow trout have usually severe disturbances in the development of sperm ducts. Therefore sperm cannot be stripped from reversed males. When using them in crosses, these fish should be sacrificed (Bye and Lincoln 1986). Sperm from reversed males of rainbow trout is obtained from surgically removed testes. After extraction, sperm is diluted with physiological solution.

6.4.2. Application of Hormonal Sex Reversal for Sex Control

Hormonal sex reversal may be used for sex regulation in two ways:

1. By direct method, i.e. masculinization or feminization of all commercially reared fish by hormonal treatment.
2. By indirect method or genetic sex regulation, i.e. by crossing of preliminarily obtained sex-reversed broodstock with normal breeders.

Direct application of hormonal sex reversal is limited since in this case fish consuming steroid hormones are intended for human consumption. This method of sex regulation is used on a large scale only in tilapia culture. In many countries, all fish are fed with methyltestosterone to obtain all-male progenies. Tilapia males grow better than females and, what is also important, rearing of monosex progenies prevents uncontrolled reproduction of this quickly maturing fish. In the United States, direct masculinization of tilapia progenies by androgen treatment is not an approved technique by the U.S. Food and Drug Administration (FDA).

Genetic sex regulation is regarded as most optimal since there is no need to treat all reared fish with hormone and hormonally treated fish are not intended for human consumption. For species with female homogamety (female - XX, male - XY), which was detected in most aquaculture species, all-female progenies may be obtained by crossing sex-reversed males XX with normal females XX. By this method all-female progenies were obtained in many aquaculture and fisheries species such as rainbow trout, common carp, grass carp, yellow perch, walleye and some others. As mentioned above, it is convenient to use all-female gynogenetic progenies as material for initial experiments on sex reversal under androgen influence. In further generations portion of fish from all-female progenies should be sex-reversed again to renew stock of reversed males (Figure 6.4.1).

Figure 6.4.1. Scheme of genetic sex regulation for production of all-female progenies in species with female homogamety (XX/XY).

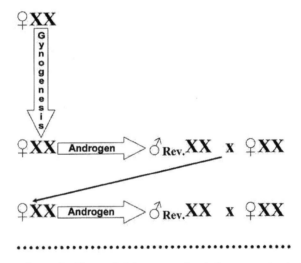

The scheme shown in Figure 6.4.1 was realized, for example, in Israel with common carp. The all-female gynogenetic progenies were produced (Cherfas et al. 1993) and used as initial material for experiments on hormonal sex reversal (Gomelsky et al. 1994). Later sex-reversed males of several consecutive generations were obtained and used for mass production of all-female progenies by their crosses with normal females. Cherfas et al. (1996) reported results of comparative rearing of all-female and normal mixed-sex progenies in climate conditions of Israel where common carp reach sexual maturity before reaching market size. It was shown that rearing of all-female progenies increased production yield by 7-8%. This increase was attributed to sexual dimorphism in body weight: females were 15% heavier than males. An additional advantage of rearing all-female progenies was prevention of uncontrolled fish reproduction in ponds.

Commercial rearing of all-female rainbow trout progenies is organized in several European countries (e.g. United Kingdom and France). It is estimated that about 80-90% of all reared fish there are all-female progenies. Females of rainbow trout have a better growth rate than males, and they do not reach sexual maturity before attaining market size. Males reach sexual maturity during rearing and this significantly decreases the quality of flesh.

The scheme for production of all-male progenies in species with female homogamety (XX/XY) is more complex and consists of several stages. The fist stage is feminization of genotypic males XY by estrogen treatment. As shown in 6.4.1 in the example with medaka, when sex-reversed females XY are crossed with normal males XY, a part of fish in the resulting progeny have genotype YY. These males, YY, when crossed with normal females XX, produce all-male progenies:

$$♂YY \times ♀XX \rightarrow 100\% ♂♂XY$$

The method of production of all-male progenies in tilapia using YY males has been elaborated by scientists from University of Wales Swansea with collaborators (Beardmore et al. 2001). At present, a commercial project for mass production of all-male progenies in tilapia is established by Fishgen Ltd. (http://www.fishgen.com). This project is called GMT[TM] or 'Genetically Male Tilapia' (as opposed to all-male tilapia produced by direct masculinization of fish). The scheme for mass production of YY males is presented in Figure 6.4.2. You can see in the figure that this program includes numerous progeny tests to identify YY males and sex-reversed females (XY and YY). Note that the last step in this scheme for mass production of YY males is the crossing YY males with sex-reversed females YY.

Figure 6.4.2. Scheme for mass production of YY male broodstock in Nile tilapia; DES is abbreviation of estrogen diethylstilbestrol (from Beardmore et al. 2001, Aquaculture 197:283-301; Copyright 2001, reproduced with permission from Elsevier).

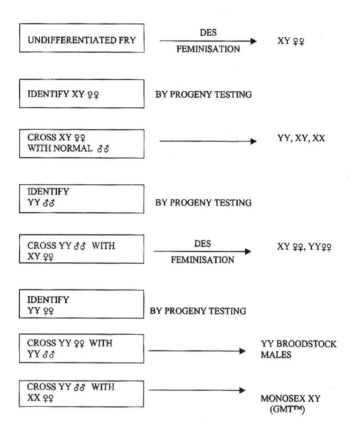

Chapter 7. Gene Engineering and Genomics

7.1. Gene Engineering and Production of Transgenic Fish

7.1.1. Introduction: Gene Engineering and Transgenic Animals

Gene engineering is a transformation of genomes by inserting a gene-containing material. Gene engineering is based on recombinant DNA technology. Recombinant DNA is DNA that has been created artificially. DNA from two or more sources is incorporated into a single recombinant molecule.

Organisms having in their genomes foreign genes are called transgenic. Introduced genes are called transgenes, i.e. transferred genes. Another name for transgenic organisms is GMO or 'genetically modified organisms'.

From the beginning, studies on gene engineering were performed only on bacteria. Later, these studies were spread to plants and animals. For the fist time transgenic mice were produced in 1982. In 1985 the first article was published on production of transgenic fish. Zhu et al. (1985) have reported creating transgenic goldfish by inserting a gene for human growth hormone. In the next several years transgenic fish were produced in several aquaculture species: rainbow trout Chourrout et al. 1986b), channel catfish (Dunham et al. 1987), and tilapia (Brem et al. 1988). During a certain period, production and investigation of transgenic fish developed rapidly and significant positive results were obtained. However, environmental and food safety concerns, negative public perception towards transgenic organisms, and the lack of permissions and regulations have prevented introducing transgenic fish as food product into commercial aquaculture.

7.1.2. Methods of Transgenic Fish Production

The basic procedure to generate transgenic fish includes the following several stages:

A. Design and synthesis of DNA construct. A certain gene is included in a vector, i.e. DNA molecule that is capable of carrying foreign genetic information. Usually bacterial plasmids are used as vectors. Transferred DNA constructs, in addition to structural genes, include regulatory sequences (promoters) that are needed for successful incorporation and expression of a construct into DNA of the host. Recombinant plasmids with the inserted gene are cloned and sufficient amounts of copies for introduction into a fish genome are produced. In the beginning, DNA constructs that contained genes and regulatory sequences taken from mammal genomes were inserted into fish genomes. Later, so called 'all-fish' DNA constructs were mainly used. All components of these constructs originated from different fish species.

B. Introduction of DNA construct to recipient. The most common way of introduction of DNA foreign material into a recipient fish is microinjection into early embryos soon after insemination. In fish it is impossible to identify visually female and male pronuclei. Therefore, DNA material is introduced into cytoplasm at the animal pole (germinal disc) where nuclear transformations during the fertilization period occur. As an illustration, Figure 7.1.1 presents microinjection of foreign DNA into germinal disc of tilapia early embryo. There is some probability that during genome DNA replication a foreign DNA construct will be inserted into the genomic DNA. To increase this probability, a large amount of DNA construct copies are injected (10^3 - 10^7 copies per embryo).

Other methods of introduction DNA foreign material have also been elaborated. Some positive results were obtained using electroporation. According to this technique the fertilized eggs are placed in a buffer solution containing foreign DNA and electric pulses are applied. Under influence of electric pulses the permeability of egg membrane increases, allowing foreign DNA to pass into the cell.

C. Screening for transgenic fish and determination of transgene expression. The next stage is the determination of success of introduction, i.e. identification of transgenic fish among all treated fish. The presence of a foreign DNA fragment in genomic DNA is determined. Among transgenic fish the presence of the gene product in tissues is checked to verify whether a transgene is active and expressed.

Figure 7.1.1. Microinjection into the germinal disc through the micropyle of a fertilized tilapia egg (from Brem et al. 1988, Aquaculture 68:209-219; Copyright 1988, reproduced with permission from Elsevier).

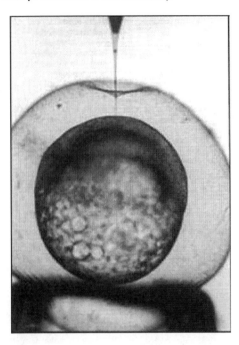

D. Study of inheritance and development of a stable line of transgenic fish. The next stage is to investigate whether the novel trait is inherited and to analyze the mode of inheritance. Injected fish may be mosaic, i.e. some cells will have an integrated fragment but some cells do not. Even in integrated cells, foreign fragments will be in heterozygote state. Therefore, it needs several generations to develop a stable line of transgenic fish. As an example, Figure 7.1.2 presents the pedigree of stable transgenic line development in medaka. Fish of this transgenic line carry DNA construct of a bacterial chloraphenicol acetyltransferase (CAT) gene and rainbow trout metallothionein-A promoter (rtMT-A). As shown in Figure 7.1.2, this transgenic line was originated from one injected mosaic transgenic female (designated in the figure as 'Female 0').

Figure 7.1.2. Establishment of stable transgenic line of medaka carrying rt-MT-A-CAT DNA construct (from Kinoshita et al. 1996, Aquaculture 143:267-276; Copyright 1996, reproduced with permission from Elsevier).

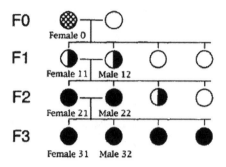

O ; wild-type, ◑ ; heterozygote for rtMT-A-CAT, ● ; homozygote for rtMT-A-CAT
⊛ ; fish in which rtMT-A-CAT integrated cells and non-integrated cells coexist

7.1.3. Properties of Transgenic Fish Valuable for Aquaculture

Up to the present, the most impressive result from the application of transgenic technology was enhancement of fish growth rate. Genes controlling synthesis of growth hormones were inserted into fish genomes most frequently. It was proposed that functioning of these transgenes would increase fish growth rate. In the first experiments genes isolated from mammals were used, later 'all-fish' DNA constructs were inserted.

Positive results were obtained after inserting a DNA construct, containing a gene for chinook salmon growth hormone linked with an antifreeze protein gene promoter from ocean pout. According to Du et al. (1992) transgenic Atlantic salmon were 2 to 6 times larger than non-transgenic control fish. Even larger differences in body weight between transgenic and non-transgenic fish were observed by Devlin et al. (1994b) when coho salmon were injected with another 'all-fish' DNA construct containing a gene for growth hormone from sockeye salmon. In this case transgenic fish were ten times heavier than non-transgenic siblings. Advantages in growth rate of transgenic fish were also revealed in common carp, tilapia and several other aquaculture species.

The first publications reporting superior growth rate were based on investigation of transgenic fish of the initial (F_0) generation. At a present, transgenic fish of further generations are produced and investigated. Some mechanisms inducing faster growth rate of transgenic fish were revealed. For example, Cook et al. (2000) have shown that transgenic Atlantic salmon of the second generation (F_2) had about 2.7-fold greater growth rate than a non-transgenic control. Increase of growth rate of transgenic fish was caused mainly by increasing fish appetite, i.e. transgenic fish consumed an artificial diet much more actively than non-transgenic fish. Devlin et al. (2000) have shown that, as a result of transgene expression, the plasma level of growth hormone in transgenic fish was 20-30 times higher than in non-transgenic fish. Some morphological abnormalities (agromegaly) indicating some imbalance in growth processes in transgenic fish were described.

Currently fast growing stable lines of transgenic Atlantic salmon, tilapia and some other aquaculture species are developed.

7.1.4. Future Perspective for Using Transgenic Fish in Aquaculture

Using transgenic fish in aquaculture, i.e. possibility to raise commercially and to sell transgenic fish as food product, is not scientific but rather an ethical and political problem. Negative perception of genetically modified organisms, and transgenic animals in particular, influences the creation and approval of corresponding regulations. Theoretically, rearing of transgenic fish could increase productivity and would benefit producers. However, taking into account negative attitudes towards genetically modified organisms in society, producers themselves may fear that "wrong labels" will repel customers.

An application to legalize rearing transgenic Atlantic salmon as a food product has been submitted by a private company to the U. S. Food and Drug Administration (FDA) about ten years ago. Up to the present, no transgenic fish have been approved as food product in the United States.

As was described above, the use of transgenic fish as food products meets with many difficulties. Other directions of transgenic fish application are of less concern. The first transgenic fluorescent aquarium fishes (medaka and zebrafish) were produced in 2001. The fish were obtained by inserting in their genomes a gene for green fluorescent protein from a jellyfish genome. These fluorescent fish glow with bright green color. A little later the fluorescent red zebrafish was produced by inserting the corresponding gene from a sea anemone genome. Transgenic fluorescent zebrafish was the first genetically modified pet, which was allowed (in 2003) to be sold in the United States. Currently these fish are pretty common in pet stores.

7.2. Fish Genomics

7.2.1. Introduction: Genomics and Its Basic Directions

Genomics is the development and application of sequencing, mapping and computational methods for the analysis of the entire genome of an organism.

Genomics is the rapidly developing branch of genetics. It appeared in connection with major advances in automated DNA sequencing and using elaborate computer programs to analyze large amounts of sequence data in the mid-1980s. The rise of genomics was directly connected with the Human Genome Project, the complex international project of many laboratories, coordinated by the U.S. Department of Energy and the National Institute of Health, to sequence the entire human genome and to identify most of its genes. Human Genome Project was started in 1990.

For DNA sequencing the Human Genome Project used so called 'map-based cloning' or a 'hierarchic shotgun' approach. This approach involves dividing the entire genomic DNA into pieces (approximately 150,000 base pairs in length) which are inserted into Bacterial Artificial Chromosome (or BAC) vectors for cloning in bacteria. The DNA fragments represented in the BAC genomic library are then organized into a physical map and individual BAC clones are selected and sequenced. Finally, the clone sequences are assembled to reconstruct the sequence of the genome.

In 1998 the private company Celera announced that it elaborated new method of genome sequencing and promised to sequence the human genome in three years. This method is called Whole Genome Shotgun Sequencing. Sequencing a genome by this method involves construction of a partially overlapping library of small (about 2,000 base pairs) genomic DNA fragments, sequencing each clone, and assembling the genomic sequence using computer program on the basis of the sequence overlaps. When compared with the hierarchic shotgun approach, the Whole Genome Shotgun approach allows the BAC step to be eliminated during the sequencing process.

Publicly supported Human Genome Project and the private company Celera simultaneously published a draft of the human genome in February 2001 issues of 'Nature' and 'Science', respectively.

Genomics has three main subfields:

1. **Structural genomics**
2. **Functional genomics**
3. **Comparative genomics**

Structural genomics involves genetic mapping, physical mapping and sequencing of entire genomes. As shown in Chapter 4, genetic maps are constructed based on recombination frequencies.

Genetic mapping determines the linkage groups and recombination distances between markers or genes but it does not indicate real locations on chromosomes. The creation of genetic maps with markers spaced at short intervals, so-called high-density genetic maps, is an important task of structural genomics. Different DNA markers are used for genetics mapping (microsatellites, AFLPs, RAPDs etc.) but Single Nucleotide Polymorphism markers (SNPs) become most important for creation of high-density genetic maps. As noted in the corresponding part of Chapter 4, identification of SNPs is amenable to automated analysis.

Currently in human genome as well as in genomes of major agricultural animals several millions of SNPs are identified.

Physical mapping determines the location of genetic markers or genes by analyzing genomic DNA directly, regardless of inheritance mode and recombination data. A high resolution physical map is a clone contig map ('contig' is short for 'contiguous'), a set of ordered, partially overlapping clones (for example, BAC) of a chromosome without any gaps. As was mentioned above, systematic sequencing of clones from contig physical maps results in sequencing of the entire genome. Actually, the complete nucleotide sequence of the chromosome is the physical map with highest resolution.

The ultimate purpose of structural genomics is the creation of integrated chromosome maps which combine information contained in genetic maps and physical maps. There are several effective methods of analysis which make it possible 'to anchor' recombinant genetic markers or genes on physical chromosome maps.

One specific feature of genomic projects is the creation of databases, which are available to research communities and the public. There are several web sites which contain information on the human genome as well as genomes of many other organisms. For example, National Center for Biotechnology Information (http://www.ncbi.nlm.nih.gov/) is a U.S. government-funded national resource for molecular biology information. This web site contains access to many public databases, including GenBank (http://www.ncbi.nlm.nih.gov/genbank/), an annotated collection of all publicly available DNA sequences.

The main purpose of **functional genomics** is the comprehensive analysis of the functions of genes in entire genomes.

The functioning genes, which code proteins, make up only of a fraction of the entire genome. The effective tool for identification of expressed genes is the synthesis and analysis of complementary DNA (or cDNA). The complementary DNA is synthesized from messenger RNA (mRNA) using a special enzyme called reverse transcriptase. Since complementary DNA is a DNA copy of mRNA, it corresponds to functional protein-coding genes in the genome. The cDNA library that contains hybrid vectors with cDNA insert may be created. The short sequences identified from cDNA are called expressed sequence tags (ESTs). Information on ESTs in genomes is accumulated in the publicly available Expressed Sequence Tags database. These data are very important for identification of functioning genes in genomes, analysis of their expression and physical mapping of genes. The EST databases for human genome currently contain more than 8 million entries.

The break-through technology which is developed and used in functional genomics for the expression analysis of thousands of genes simultaneously is DNA microarrays (or gene chips). DNA microarray is a set of DNA sequences attached to a solid glass or plastic slide in an orderly manner. Each spot on the cDNA microarray contains a particular sequence of cDNA isolated from the organism. The first step of a comparative hybridization experiment using cDNA microarray for gene expression analysis involves isolation of mRNA from two separate samples (for example, samples from different tissues or samples taken from healthy and sick individuals). Then fluorescently labeled cDNAs are synthesized. A mixture of cDNAs is hybridized to the DNA microarray containing the full set of cDNA sequences. Then microarray is scanned using a special fluorimager, and the color of each spot is determined. Genes expressed only in first sample would be red in color, genes expressed only in the second sample would be green and those genes expressed equally in both samples would be yellow. This allows researchers to determine specific genes expressed in different tissues or genes that are functioning in response to disease. (cDNA microarrays is one type of DNA microarrays. The other type is oligonucleotide DNA microarrays which can be used for mass identification of Single Nucleotide Polymorphism markers, SNPs, in genome.)

Comparative genomics is the comparison of entire genomes of different species, with the goal of enhancing understanding of the functions and evolutionary relationships of each genome. In addition to the human genome sequencing in the Human Genome Project, a comparative genome sequencing of 8 model species representing different organism groups was performed (including fugu or puffer fish,

Fugu rubripes). The information on genomics of this model species and other model fish species is given in the next part of this chapter (7.2.2).

7.2.2. Current Status of Fish Genomics

The genomic studies of fish model species are supported by funds for the fundamental and health sciences. As was mentioned in previous part, fugu (or puffer fish) was chosen as a model species for Human Genome Project. The sequencing of the fugu genome was completed in 2002. Fugu was chosen as a model species since its genome is about eight times smaller than that of humans, but with a similar estimated number of genes. The smaller gene density makes puffer fish DNA much easier to work with than human DNA. Since many of the fugu genes are homologous to human genes, once genes are identified in the puffer fish genome their homologies might be found in the human genome. This method for gene searching is widely used in comparative genomics.

Other three model fish species whose genomes have been sequenced up to now are zebrafish, Tetraodon and medaka. Zebrafish *(Danio rerio)* is a major model species for investigation of vertebrate development and gene functioning. The zebrafish genome has been sequenced and actively analyzed by joint effort between scientific groups in the United States, United Kingdom and some other countries. The detailed chromosome maps are compiled for this model species; the large amount of SNPs has been identified and mapped. The number of deposited ESTs from the zebrafish genome is more than 1.5 million entries.

In the United States, the studies on genomics of fish non-model species, first of all of major aquaculture species, is coordinated by National Aquaculture Genome Projects (http://www.animalgenome.org/aquaculture/) as a part of the National Animal Genome Research Program (NAGRP) (http://www.animalgenome.org/) at the United States Department of Agriculture. The aquaculture species (or groups of species) within the NAGRP initiative are catfish, salmonids, tilapia, striped bass and oysters. The studies on genomics of some aquaculture species are performed in several other countries. The scientific groups working with the same or close species collaborate and share open genomic databases.

The main directions of genomic studies for aquaculture species are listed below.

In the area of structural genomics:

• Development of high density genetic linkage maps
• Development of BAC libraries
• Development of BAC-based physical maps
• Merging genetic linkage maps with physical maps to produce integrated maps
• Partial or complete genome sequencing

In the area of functional genomics:

• Creation of ESTs collections
• Development of cDNA microarrays

As you can see, these directions of genomic studies are the same ones, which were realized in studies of human genome and genomes of model species. In recent several years a significant progress in genomics of major aquaculture species has been made. Below some achievements in genomics of several aquaculture species are listed.

Significant genome resources have been developed for channel catfish and channel catfish x blue catfish hybrids, fish which account more than 60% of the U.S. aquaculture production. According to Lu et al. (2010) these resources include over a half million expressed sequenced tags (ESTs), a large number of genome sequences generated from BAC, genetic linkage maps, genome physical maps, tens of thousands of microsatellite markers, hundreds of thousands SNPs, over 10,000 full-length cDNAs. The United States Department of Agriculture has provided funding for catfish whole-genome sequencing and development of high-density SNP microarrays for catfish. In order to accumulate genome resources for catfish recently a catfish genome database, cBARBEL, the Catfish Breeder and Researcher Bioinformatics Entry Location (www.catfishgenome.org), was established.

Considerable genomic resources have been developed over the past decade for Atlantic salmon. According to (Davidson et al. 2010) these resources include approximately 500,000 ESTs and a catalog of genes, a BAC library and physical map, a linkage map that is integrated with the physical map and karyotype, an expression array that can survey 32,000 transcripts, as well as a SNP chip that can detect 5,600 genetic loci. International Collaboration to Sequence the Atlantic Salmon Genome (ICSASG) was established. The project on sequencing Atlantic salmon entire genome was started in 2010.

Considerable genomic resources have been developed for several other major aquaculture species such as rainbow trout and tilapia. Currently project on tilapia genome sequencing is being performed by Broad Institute, genomic medical research center in Cambridge, Massachusetts www.broadinstitute.org/models/tilapia.

The genomic projects require significant investments and combined efforts of many scientific groups. The current stage of fish genomics mostly involves accumulation of genomic resources. But in the future, based on this information, the new advanced methods of genetic improvement will be elaborated. In Chapter 4 such methods as identification of Quantitative Trait Loci and Marker Assisted Selection have been described. These are the initial aspects of possible practical applications of genomic studies. New specialized branches of genomics (so-called "omics" disciplines) are currently being created. For example, **proteomics** analyzes the function and structure of all proteins controlled by organism's genome. **Pharmacogenomics** is directed to creation of medicines which are customized to person's genetic profile. **Nutrigenomics** investigates how different foods may interact with organism's specific genes. There are predictions that further development of genomics may revolutionize the future of medicine, food and animal sciences. Apparently a similar revolution will occur in fish genetics.

References

Agresti, J.J., Seki, S., Cnaani, A., Poompuang, S., Hallerman, E.M., Umiel, N.,Hulata, G., Gall, G.A.E., and May, B. 2000. Breeding new strains of tilapia: development of an artificial center of origin and linkage map based on AFLP and microsatellite loci. Aquaculture 185:43-56.

Alsaqufi, A.S. 2011. Application of microsatellite DNA markers in studies on induced gynogenesis in ornamental (koi) carp (*Cyprinus carpio* L.). Master's Thesis. Kentucky State University.

Anderson E.C. and Garza, J.C. 2006. The power of single nucleotide polymorphisms for large scale parentage analysis. Genetics 172: 2567-2582.

Barman, H.K., Barat, A., Yadav, B.M., Banerjee, S., Meher, P.K., Reddy, P.V.G.K and Jana, R.K. 2003. Genetic variation between four species of Indian major carps as revealed by random amplified polymorphic DNA assay. Aquaculture 217:115-123.

Baroiller, J.F., Clota, F., and Geraz, E. 1995. Temperature sex determination in two tilapia, *Oreochromis niloticus* and the red tilapia (red Florida strain): effect of high and low temperature. In: Proc. of Fifth Intern. Symp. Reproduct. Physiol. of Fish, p.158-160.

Beardmore, J.A., Mair, G.C., and Lewis, R.I. 2001. Monosex male production in finfish as exemplified by tilapia: applications, problems, and prospects. Aquaculture 197:283-301.

Beck, M.L., Biggers, C.J., and Dupree, H.K. 1980. Karyological analysis of *Ctenopharyngodon idella*, *Arichthys nobilis*, and their F1 hybrid. Trans. Amer. Fish. Soc. 109:433-438.

Bondari, K. and Dunham, R. A. 1987. Effects of inbreeding on economic traits of channel catfish. Theor. Appl. Genet. 74:1-9.

Bosworth, B.G., Silverstein, J.T., Wolters, W.R., Li, M.H., Robinson, E.H. 2007. Family, strain, gender, and dietary protein effects on production and processing traits of Norris and NWAC103 strains of channel catfish. North Amer. J. Aquac. 69:106-115.

Brem, G., Brenig, B., Hörstgen-Schwark, G. and Winnacker, E.L. 1988. Gene transfer in tilapia (Oreochromis niloticus). Aquaculture 68:209-219.

Bridges, W.R. and von Limbach, B. Inheritance of albinism in rainbow trout. 1972. J. of Heredity 63:152-153.

Bye, V.J., and Lincoln, R.F. 1986. Commercial methods for the control of sexual maturation in rainbow trout (*Salmo gairdneri* R.). Aquaculture 57:299-309.

Cassani, J.R., Caton, W.E. 1986. Efficient production of triploid grass carp (*Ctenopharyngodon idella*) utilizing hydrostatic pressure. Aquaculture 46:37-44.

Cherfas, N.B., Gomelsky, B., Peretz, Y., Ben-Dom, N., Hulata, G. and Moaz, B. 1993. Induced gynogenesis and polyploidy in the Israeli common carp line Dor-70. Isr. J. Aquac. – Bamidgeh 45:59-72.

Cherfas, N.B., Gomelsky, B., Ben-Dom, N., Joseph, D., Cohen S., Israel, I., Kabessa, M., Zohar, G., Peretz, Y., Mires, D. and Hulata, G. 1996. Assessment of all-female common carp progenies for fish culture. Isr. J. Aquac.- Bamidgeh 48:149-157.

Cherfas, N.B., Gomelsky, B.I., Emelyanova, O.V. and Recoubratsky, A.V. 1994. Induced diploid gynogenesis and polyploidy in crucian carp, *Carassius auratus gibelio* (Bloch), x common carp, *Cyprinus carpio* L., hybrids. Aquacult. Fish. Manag. 25:943-954.

Chourrout, D. 1982. Gynogenesis caused by ultraviolet irradiation of salmonid sperm. J. Exp. Zool. 223:175-181.

Chourrout, D. 1984. Pressure-induced retention of second polar body and suppression of first cleavage in rainbow trout: production of all-triploids, all-tetraploids, and heterozygous and homozygous diploid gynogenetics. Aquaculture 36:111-126.

Chourrout, D., Chevassus, B., Krieg, F., Happe, A., Burger, G., and Renald, P. 1986a. Production of second generation triploid and tetraploid rainbow trout by mating tetraploid males and diploid females – potential of tetraploid fish. Theor. Appl. Genet. 72:193-206.

Chourrout, D., Guyomard, R. and Houdebine L. 1986b. High efficiency gene transfer in rainbow trout (*Salmo gairdneri* Rich.) by microinjection into egg cytoplasm. Aquaculture 51:143-150.

Cook, J.T., McNiven, M.A., Richardson, G.F., and Sutterlin, A.M. 2000. Growth rate, body composition and feed digestibility / conversion of growth-enhanced transgenic Atlantic salmon (*Salmo salar*). Aquaculture 188:15-32.

Clark, F.H. 1970. Pleiotropic effects of the gene for golden color in rainbow trout. J. of Heredity 61:8-10.

Crandell, P.A. and Gall, G.A.E. 1993. The genetics of body weight and its effect on early maturity based on individually tagged rainbow trout (*Oncorhynchus mykiss*) Aquaculture 117:77-93.

Cuevas-Uribe, R., Gomelsky, B., Mims, S.D. and Pomper, K.W. 2009. Progress in studies on hormonal sex reversal and genetic sex control in black crappie. Rev. Fish. Sci. 17:1-7.

Dabrowski, K., Rinchard, J., Lin, F., Garcia-Abiado, M. and Schmidt, D. 2000. Induction of gynogenesis in muskellunge with irradiated sperm of yellow perch proves diploid muskellunge male homogamety. J. Exp. Zool. 287:96-105.

Davidson, W.S., Koop, B.F., Jones, S.J.M., Iturra, P., Vidal, R., Maass, A.,Jonassen, I., Lien, S. and Omholt, S.W. 2010. Sequences the genome of the Atlantic salmon (*Salmo salar*). Genome Biology 11:403.

Dawley, R.M., Graham, J.H. and Schultz, R.J. 1985. Triploid progeny of pumpkinseed x green sunfish hybrids. J. Hered. 76:251-257.

Devlin, R.H., Mc Neil, B.K., Solar, I.I., and Donadldson, E.M. 1994a. A rapid PCR based test for Y-chromosomal DNA allows simple production of all-female strains of Chinook salmon. Aquaculture 128:211-220.

Devlin, R.H., Yesaki, T.Y., Biagi, C.A., Donaldson, E.M., Swanson, P., and Chen, W.K. 1994b. Extraordinary salmon growth. Nature 371:209-210.

Devlin, R.H., Swanson, P., Clarke, W.C., Plisetskaya, E., Dickoff, W., Moriyama, S., Yesaki, T.Y., and Hew, C.L. 2000. Seawater adaptability and hormone levels in growth-enhanced transgenic coho salmo, *Oncorhynchus kisutch*. Aquaculture 191:367-385.

Devlin, R.H. and Nagahama, Y. 2002. Sex determination and sex differentiation in fish: an overview of genetic, physiological, and environmental influences. Aquaculture 208:191-364.

Dobosz, S., Goryczko, K., Kohlmann, K., and Korwin-Kossakowski, M. 1999. The yellow color inheritance in rainbow trout. J. of Heredity 90:312-315.

Du, S.J., Gong, Z., Fletcher, G.L., Shears, M.A., King, M.J., Idler, D.R., and Hew, C.L. 1992. Growth enhancement of transgenic Atlantic salmon by the use of an "all-fish" chimeric growth hormone gene construct. Bio/Technology 10:176-181.

Dunham, R.A. and Smitherman, R.O. 1983. Response to selection and realized heritability for body weight in three strains of channel catfish, *Ictalurus punctatus*, grown in earthen ponds. Aquaculture 33:89-96.

Dunham, R.A., Eash, J., Askins, J. and Townes, T.M. 1987. Transfer of the metallothionein-human growth hormone fusion gene into channel catfish. Trans. Amer. Fish. Soc. 116:87-91.

Ehlinger, N.F. 1977. Selective breeding of trout for resistance to furunculosis. NY Fish Game J. 24:25-36.

Evans, F., Matson, S., Brake, J. and Langdon, C. 2004. The effects of inbreeding on performance traits of adult Pacific oysters (*Crassostrea gigas*). Aquaculture 230:89-98

Falconer, D.S. and Mackay, T.F.C. 1996. Introduction to quantitative genetics. Fourth edition. Prentice Hall.

Flajšans, M. and Ráb, P. 1990. Chromosome study of *Oncorhynchus mykiss* kamloops. Aquaculture 89:1-8.

Funkenstein, B., Cavari, B., Stadie, T. and Davidovitch (Yaiche), E. 1990. Restriction site polymorphism of mitochondrial DNA of the gilthead sea bream (*Sparus aurata*) broodstock in Eilat, Israel. Aquaculture 89:217-223.

Garcia de Leon, F.J., Canonne, M., Quillet, E., Bonhomme, F., and Chatain, B. 1998. The application of microsatellite markers to breeding programmes in the sea bass, *Dicentrarchus labrax*. Aquaculture 159:303-316.

Gjedrem, T. 2000. Genetic improvement of cold-water fish species. Aquaculture Research 31:25-33.

Gjedrem, T. and Baranski, M. 2009. Selective breeding in aquaculture: an introduction. Springer.

Goetz, F.W., Donaldson, E.M., Hunter, G.A. and Dye, H.M. 1979. Effects of estradiol-17β and 17α-methyltestosterone on gonadal differentiation in the coho salmon, *Oncorhynchus kisutch*. Aquaculture 17:267-278.

Gomelsky, B. 1985. Hormonal sex inversion in common carp (*Cyprinus carpio* L.). Ontogenez 16:398-405 (in Russian with English summary).

Gomelsky, B., Cherfas, N.B., Peretz, Y., Ben-Dom, N., and Hulata, G. 1994. Hormonal sex inversion in the common carp. Aquaculture 126:265-270.

Gomelsky, B., Mims, S.D., Onders, R.J., Shelton, W.L., Dabrowski, K., and Garcia-Abiado, M.A. 2000. Induced gynogenesis in black crappie. North Amer. J. of Aquac. 62:33-41.

Gomelsky, B., Cherfas, N.B., Ben-Dom, N. and Hulata, G. 1996. Color inheritance in ornamental (koi) carp (*Cyprinus carpio* L.) inferred from color variability in normal and gynogenetic progenies. Isr. J. Aquac. – Bamidgeh 48:219-230.

Gomelsky, B., Cherfas, N.B. and Hulata, G. 1998. Studies on the inheritance of black patches in ornamental (koi) carp. Isr. J. Aquac. – Bamidgeh 50:134-139.

Gomelsky, B., Mims, S.D., Onders, R.J. and Novelo, N.D. 2005. Inheritance of predorsal black stripe in black crappie. North Amer. J. Aquac. 67:167-170.

Gomelsky, B., Schneider, K.J., and Alsaqufi, A.S. 2011. Inheritance of long fins in ornamental koi carp. North Amer. J. of Aquac. 73:1-4.

Hara, M. and Sekino, M. 2003. Efficient detection of parentage in a cultured Japanese flounder *Paralichthys olivaceus* using microsatellite DNA marker. Aquaculture 217:107-114.

Hershberger, W.K., Myers, J.M., Iwamoto, R.N., Mcauley, W.C. and Saxton, A.M. 1990. Genetic changes in the growth of coho salmon (*Oncorhynchus kisutch*) in marine net-pens, produced by ten years of selection. Aquaculture, 85:187-197.

Hörstgen-Schawrk, G. 1993. Selection experiments for improving "pan-size" body weigh of rainbow trout (*Oncorhynchus mykiss*). Aquaculture 112:13-24.

Hunter, G.A., Donaldson, E.M., Stoss, J. and Baker, I. 1983. Production of monosex female groups of Chinook salmon (*Oncorhynchus tschwytscha*) by the fertilization of normal ova with sperm from sex-reversed females. Aquaculture 33:355-364.

Hussain, M.G., Islam, M.S., Hossain, M.A., Wahid, M.I., Kohinoor, A.H.M., Dey, M.M., and Mazid, M.A. 2002. Stock improvement of silver barb (*Barbodes gonionotus* Bleeker) through several generations of genetic selection. Aquaculture 204:469-480.

Ibarra, A.M., Ramirez, J.L., Ruiz, C.A., Cruz, P., and Avila, S. 1999. Realized heritabilities and genetic correlation after dual selection for total weight and shell width in catarina scallop (*Argopecten venricosus*). Aquaculture 175:227-241.

Ilyasov, Ju.I., Kirpichnikov, V.S. and Shart, L.A. 1983. Methods and effectiveness of common carp selection for higher resistance to dropsy. In: Biological basics of fish culture: problems of genetics and selection, Nauka, p. 130-146 (in Russian).

Jensen, G.L., Shelton, W.L., Yang, S. and Wilken, L.O. 1983. Sex reversal of gynogenetic grass carp by implantation of methyltestosterone. Trans. Amer. Fish. Soc. 112:79-85.

Johnson, K.R. and Wright, J.E. 1986. Female brown trout x Atlantic salmon hybrids produce gynogens and triploids when backcrossed to male Atlantic salmon. Aquaculture 57:345-358.

Johnstone, R., Simpson, T.H. and Youngson, A.F., 1978. Sex reversal in salmonid culture. Aquaculture, 13: 115-134.

Katasonov, V.Y. 1978. A study of pigmentation in hybrids between the common and decorative Japanese carp. III, The inheritance of blue and orange patterns of pigmentation. Genetika 14:2184-2192 (in Russian with English summary).

Kato, K., Hayashi, R., Yuasa, D., Yamamoto, S., Miyashita, S., Murata, O. and Kumai, H. 2002. Production of cloned red sea bream, *Pagrus major*, by chromosome manipulation. Aquaculture 207:19-27.

Khaw, H.L., Ponzoni, R.W. and Danting, M.J.C. 2008. Estimation of genetic change in the GIFT strain of Nile tilapia (*Oreochromis niloticus*) by comparing contemporary progeny produced by males born in 1991 or in 2003. Aquaculture 275:64-69.

Kincaid, H.L. 1976. Inbreeding in rainbow trout (*Salmo gairdneri*). J. Fish. Res. Board Can. 33:2420-2426.

Kincaid, H.L. 1977. Rotational line crossing: an approach to the reduction of inbreeding accumulation in trout brood stocks. Progr. Fish Cult. 39:179-181.

Kirpichnikov, V.S. 1981. Genetic bases of fish selection. Springer-Verlag.

Kirpichnikov, V.S., Ilysov, Ju.l., Shart, L.A., Vikhman, A.A., Ganchenko, A.L., Ostashevsky, A.L., Simonov, V.M., Tikhonov, G.F. and Tjurin, V.V. 1993. Selection of Krasnodar common carp (*Cyprinus carpio* L.) for resistance to dropsy: principal results and prospects. Aquaculture 111:7-20.

Knibb, W., Gorshkova, G. and Gorshkov, S. 1998. Genetic improvement of cultured marine finfish: case studies. In: Tropical mariculture, Academic Press, p. 111-149.

Kinoshita, M., Toyohara, H., Sakaguchi, M., Inoue, K., Yamashita, S., Satake, M., Wakamatsu, Y., and Kenjiro, O. 1996. A stable line of transgenic medaka (*Oryzias latipes*) carrying the CAT gene. Aquaculture 143:267-276.

LeFever, W. 1991. The new butterfly koi. Tropical Fish Hobbyist, November: 78-83.

LeFever, R., 2010. The origin of butterfly koi. Pond Trade Magazine March/April: 12-14.

Li, M.H., Manning, B.B., Robinson, E.H., Bosworth, B.G. and Wolters, W.R. 2001. Comparison of growth, processing yield, and body composition of USDA 103 and Mississippi "normal" strains of channel catfish fed diet containing three concentrations of protein. J. Word Aquac. Soc. 32:402-408.

Liu, Z., Li, P., Kucuktas, H., Nichols, A., Tan, G., Zheng, X., Argue, B.J., Dunham, R.A. and Yant, D.R. 1999. Development of amplified fragment length polymorphism (AFLP) markers suitable for genetic linkage mapping of catfish. Trans. Amer. Fish. Soc. 128: 317-327.

Longalong, F.M., Eknath, A.E. and Bentsen, H.B. 1999. Response to bi-directional selection for frequency of early maturing females in Nile tilapia (*Oreochromis niloticus*). Aquaculture 178: 13-25.

Lu, J., Peatman, E., Yang, Q., Wang, S., Hu, Z., Reecy, J., Kucuktas, H., and Liu, Z. 2011. The catfish genome database cBARBEL: an informatic platform for genome biology of ictalurid catfish. Nucleic Acids Research 39 (suppl. 1):D815-D821.

Lutz, C.G. 1999. The Nile pearl: an unusual color morph in *Oreochromis niloticus*. Aquaculture Magazine 25(6):74-77.

Malison, J.M., Kayes, T.B., Best, C.D., Amundson, C.H., and Wentworth, B.C. 1986. Sexual differentiation and use of hormones to control sex in yellow perch (*Perca flavescens*). Can. J. Fish. Aquat. Sci. 43:26-35.

McAndrew, B.J., Roubal, F.R., Roberts, R.J., Bullock, A.M. and McEwen, I.M. 1988. The genetics and histology of red blond and associated variants in *O. niloticus*. Genetica 76:127-137.

Mia, M.Y., Taggart, J.B., Gilmour, A.E., Gheyas, A.A., Das, T.K., Kohinoor, A.H.M., Rahman, M.A., Sattar, M.A., Hussain, M.G., Mazid, M.A., Penman, D.J. and McAndrew, B.J. 2005. Detection of hybridization between Chinese carp species (*Hypophthalmichthys molitrix* and *Aristichthys nobilis*) in hatchery broodstock in Bangladesh, using DNA microsatellite loci. Aquaculture 247:267-273.

Nakamura, M., 1981. Feminization of masu salmon, *Oncorhynchus masou* by administration of estradiol-17β. Bull. Jpn. Soc. Sci. Fish. 47:1529.

Nakamura, N. and Kasahara, S. 1956. A study on the phenomenon of tobi koi or shoot carp. II. On the effect of particle size and quality of the food. Nippon Suisan Gakkaishi 21:1022-1024 (in Japanese with English summary). English translation is published in Isr. J. Aquac. – Bamidgeh 1977 29:44-47.

Nakamura,M. and Takahashi, H., 1973. Gonadal sex differentiation in *Tilapia mossambica* with special regard to the time of estrogen treatment of effective inducing complete feminization of genetic males. Bull. Fac. Fish. Hokkaido Univ. 24:1-13.

Neff, B.D. 2001. Genetic paternity analysis and breeding success in bluegill sunfish (*Lepomis macrochirus*). J. of Heredity 92:111-119.

Novelo, N.D. 2008. Variability and inheritance of white-red color complex and random amplified polymorphic DNA (RAPD) markers in ornamental (koi) carp. Master's Thesis, Kentucky State University.

Okutsu T, Shikina S, Kanno M, Takeuchi Y, and Yoshizaki G. 2007. Production of trout offspring from triploid salmon parents. Science 317: 1517.

Palti, Y., Parsons, J. and Thorgaard, G.H. 1997. Assessment of genetic variability among strains of rainbow and cutthroat trout using multilocus DNA fingerprints. Aquaculture 149:47-56.

Peruzzi, S., Chatain, B., Saillant, E., Haffray, P., Menu ,B., and Falguiere, J. 2004. Production of meiotic gynogenetic and triploid sea bass, *Dicentrarachus labrax* L. 1. Performances, maturation and carcass quality. Aquaculture 230:41-64

Postlethwait, J., Johnson, S., Midson, C.N., Talbot, W.S., Gates, M., Ballinger, E.W., Africa, D., Andrews, R., Carl, T., Eison, J.S., Horne, S., Kimmel, C.B. et al. 1994. A genetic linkage map for the zebrafish. Science 264:699-703.

Quinton, C.D., McMillan, I. and Glebe, B.D. 2005. Development of an Atlantic salmon (*Salmo salar*) genetic improvement program: genetic parameters of harvest body weight and carcass quality traits estimated with animal models. Aquaculture 257:211-217.

Reagan, R.E., Pardue, G.B. and Eisen, E.J. 1976. Predicting selection response for growth of channel catfish. J. of Heredity 67:49-53.

Recoubratsky, A.V., Gomelsky, B.I., Emelyanova, O.V., and Pankratyeva, E.V. 1992. Triploid common carp produced by heat shock with industrial fish-farm technology. Aquaculture 108:13-19.

Reich, L., Don, J., and Avtalion, R.R. 1990. Inheritance of the red color in tilapias. Genetics 80:195-200.

Rothbard, S. and Wohlfarth, G. 1993. Inheritance of albinism in the grass carp, Ctenopharyngodon idella. Aquaculture 115:253-271.

Sakamoto, T., Danzmann, R.G., Okamoto, N., Ferguson, M.M. and Ihssen, P.E. 1999. Linkage analysis of quantitative trait loci associated with spawning time in rainbow trout (Oncorhynchus mykiss). Aquaculture 173:33-43.

Schneider, K. J. 2010. Genetic diversity of cultured and wild populations of the freshwater prawn Macrobrachium rosenbergii based on microsatellite analysis. Master's Thesis. Kentucky State University.

Schröder, J.H. 1976. Genetics for aquarists. TFH Publications.

Sekino, M. and Hara, M. 2001. Inheritance of microsatellite DNA loci in experimental families of Japanese flounder Paralichthys olivaceus. Mar. Biotechnol. 3:310-315.

Siitonen, L., and Gall, G.A.E. 1989. Response to selection for early spawn date in rainbow trout, Salmo gairdneri. 1989. Aquaculture 78:153-161.

Smith, C.T., Seeb, J.E., Schwenke, P. and Seeb, L.W. 2005. Use of the 5'-nuclease reaction for single nucleotide polymorphism genotyping in Chinook salmon. Trans. Amer. Fish. Soc. 134:207-217.

Solar, I.I., Donaldson, E.M., and Hunter, G.A. 1984. Induction of triploidy in rainbow trout (Salmo gairdneri) by heat shock, and investigation of early growth. Aquaculture 42:57-67.

Storcet, A., Strand, C. Wetten, M., Kjøglum, S. and Ramstad, A. 2007. Response to selection for resistance against infectious pancreatic necrosis in Atlantic salmon (Salmo salar L.).

Streisinger, G., Walker, C., Dower, N., Knauber, D., and Singer, F. 1981. Production of clones of homozygous diploid zebra fish (Brachydanio rerio). Nature 291:293-296.

Tan, J.C.S., and Phang, V.P.E. 1994. Fin length inheritance in Brachydanio rerio. J. of Heredity 85:415-416.

Taniguchi, N., Takagi, M. and Mikita, K. 1999. Microsatellite DNA markers for monitoring genetic change in hatchery stocks of red sea bream (Pagrus major): a case study. In: Genetics in sustainable fisheries management, p. 206-218, Blackwell Science.

Tayamen, M.M. and Shelton, W.L. 1968. Inducement of sex reversal in Sarotherodon niloticus (Linnaeus). Aquaculture 14:349-354.

Thodesen, J., Grisdale-Helland, B., Helland, S.J. and Gjerde, B. 1999. Feed intake, growth and feed utilization of offspring from wild and selected Atlantic salmon (*Salmo salar*). Aquaculture 180:237-246.

Thodesen, J., Gjerde, B., Grisdale-Helland, B. and Storebakken, T. 2001. Genetic variation in feed intake, growth and feed utilization in Atlantic salmon (*Salmo salar*). Aquaculture 194:273-281.

Thodesen, J. and Gjedrem, T. 2006. Breeding programs on Atlantic salmon in Norway – lesson learned. In: Development of aquatic animal genetic improvement and dissemination programs: current status and action plans, WorldFish Center, p. 22-26

Thorgaard, G.H. 1977. Heteromorphic sex chromosomes in male rainbow trout. Science 196: 900-902.

Thorgaard, G.H. 1978. Sex chromosomes in the sockeye salmon: a Y-autosome fusion. Can. J. Genet. Cytol. 20:349-54.

Thorgaard, G.H., Spruell, P., Wheeler, P.A., Scheerer, P.D., Peek, A.S., Valentine, J.J. and Hilton, B. 1995. Incidence of albinos as a monitor for induced triploidy in rainbow trout. Aquaculture 137:121-130.

Van Eenennaam, A.L., Murray, J.D., and Medrano, J.F. 1999. Karyotype of the American green sturgeon. Trans. Amer. Fish. Soc. 128: 175-177.

Vasilyev, V.P., Makeeva, A.P. and Ryabov, I.N. 1975. On the triploidy of distant hybrids of carp (Cyprinus carpio) with other representativres of Cyprinidae family. Genetika 11:49-65 (in Russian with English summary).

Volckaert, F.A.M. and Hellemans, B. 1999. Survival, growth and selection in a communally reared multifactorial cross of African catfish (*Clarias gariepinus*). Aquaculture 171:49-64.

Waldbieser, G.C. and Wolters, W.R. 1999. Application of polymorphic microsatellite loci in a channel catfish *Ictalurus punctatus* breeding program. J. World Aquac. Soc. 30:256-262.

Wattendorf, R.J. 1986. Rapid identification of triploid grass carp with a Coulter Counter and Channelyzer. Progr. Fish-Cult. 48:125-132.

Wilkens, H. 1971. Genetic interpretation of regressive evolutionary processes: studies on hybrid eyes of two *Astyanax* cave populations (Characidae, Pisces). Evolution 25:530-544.

Wilkens, H. 1988. Evolution of genetics of epigean and cave *Astyanax fasciatus* (Characidae, Pisces). Evol. Biol. 23:271-367.

Wolters, W.R., Libey, G.S., and Chrisman C.L. Induction of triploidy in channel catfish. 1981. Trans. Amer.Fish. Soc. 110: 310-312.

Wohlfarth, G.W., Rothbard, S., Hulata, G. and Sweigman, D. 1990. Inheritance of red body coloration in Taiwanese tilapias and in *Oreochromis mossambicus*. Aquaculture 84:219-234.

Wohlfarth G. and Moav, R. 1972. The regression of weight gain on initial weight in carp. I. Methods and results. Aquaculture 1:7-28.

Wright, J.E. 1972. The palomino rainbow trout. PA Angler Mag. 41:8-9.

Yamamoto, T. 1969. Sex differentiation. In: Fish Physiology, v. III, Academic Press, p. 117-175.

Zhu, Z., He, L., and Chen, S. 1985. Novel gene transfer into the fertilized eggs of goldfish (*Carassius auratus* L. 1758). J. Appl. Ichthyol. 1:31-34.